Bob-Lo

AN ISLAND IN TROUBLED WATERS

~~~~~~~~~~~~~~~~~~~~~~~~~~~~~~~~~~~~

by ANNESSA CARLISLE

MOMENTUM BOOKS, L.L.C.
ROYAL OAK, MICHIGAN

Published by Momentum Books, L.L.C.

117 West Third Street
Royal Oak, Michigan 48067
www.momentumbooks.com

ISBN-13: 978-1-879094-75-8
ISBN-10: 1-879094-75-4
LCCN: 2005930179

DEDICATION

This book is dedicated to Dorothy Tresness—to her independent spirit, her passion for life and her conviction to stand up for what is rightfully hers. I also dedicate this to Michael, for reasons too numerous to list, and to Austin and Bryce, who although are too young to remember Bob-Lo, are now its biggest fans.

ACKNOWLEDGEMENTS

A big thank-you first and foremost to my mom, Linda Ball,
for helping me find the story, and to all family members who
encouraged and supported me. Also thanks to Sherry Westphal and
Russell Deline for trusting me; to Doris Guffey for sharing her memories
of Tresness and the idyllic Bob-Lo; and to the numerous collectors
and enthusiasts I met along the way who helped in the gathering
of facts and artifacts from the playground in the river.

Special appreciation to Dr. David Miramonti, whose knowledge
of the island and beloved boats is unfathomable, and whose generosity
is unlimited. For it is much of his personal collection that brings
Bob-Lo to life in the amazing photo section in this book.

CONTENTS

~~~~~~

*I*NTRODUCTION

History is made up of people, not just places. And in the history of Bob-Lo Island there is no one more interesting than Dorothy Tresness.

While growing up I was vaguely aware of Dorothy. She was my grandmother's friend: The one who lived on an island. The one who occasionally showed up at family weddings and baby showers in style—by boat. But I never knew her or her story. Not until four years ago, when I wrote her a letter.

And so this book began with a letter to Dorothy, a woman I barely knew, asking if she would tell me a little bit about the island. What followed can only be described as a wonderful journey of discovery—the discovery of a great story that satisfied the journalist inside me as it continued to unfold as news from day to day, but more importantly, it was the discovery of a woman who had a lot to share. She shared with me the importance of conviction, of having a purpose, and making life interesting.

If you're reading this for the history of Bob-Lo Island, you will be satisfied. In fact, when I first started writing it and people would ask about the subject, I would reply, "It's a book about Bob-Lo Island." But the history of Bob-Lo Island cannot—and should not—be told without Dorothy's story. And now that the book is complete, my answer is different.

So, here is the book about a woman who lives on Bob-Lo Island. ...

ac

PREFACE

July 1993 –
With the bathtub three-fourths full, the old woman scurries about to find jars, milk jugs, garbage cans and shoes—anything that would hold water. Alone in a cottage on an island floating between the coasts of Amherstburg, Canada and Detroit, she had just been notified that in 20 minutes her water supply would be shut off. The directive came from the island's owner, whose primary concern was running the profitable amusement park on the south end. For reasons unknown, he had decided that day that he was no longer interested in serving as a utility company for the island's only resident.

August 2004 –
In the moonlight, on the island just off the coast of Canada and Detroit, the woman, now 83 years old, showers in her front yard. Her shower consists of a hose running from her new neighbor's house and milk jugs full of water warmed by the sun. It is 11 years later and the waters around this island are even more turbulent. The amusement park that operated for 96 years is gone. Where roller coasters, a carousel and children's rides once stood is a residential development made up of Victorian-style houses with price tags ranging from $1 - $4 million. The large homes are impressive, done in muted colors of sand and water, turrets jutting toward the sky with rows and rows of windows taking in the vantage point of the shoreline. Further inland are condominiums, done in the same neo-Victorian architectural style as the large houses. A new road separates the development from the woman's modest cottage. But the setting is deceiving. These seemingly tranquil residences are now experiencing the complexity that has defined Bob-Lo Island since its beginnings. Like a squall that suddenly rises from the water, trouble has a way of washing over this island time and time again. It is trouble that the woman, alone and forced to live without water for all these years, has weathered before.

As Bob-Lo's longest private resident, Dorothy Tresness has lived through five different owners of the island, and may live to see yet another. In 2004, the Canadian government placed the island into receivership, leaving not only Dorothy, but all residents literally marooned. The current owner had lost control and, as a result, the residents were cut off from transportation to the mainland.

TURBULENT HISTORY

Before there were private owners, whole countries fought over Bob-Lo. France, Britain, the United States and Canada have all flown their flags over it. It was once home to a British military outpost and hosted the very first engagement of the War of 1812. But it will most likely be remembered as the home of Bob-Lo Island Amusement Park, a park that started out modestly and grew to rival any major theme park of the 20th century. For nearly 100 years, the amusement park was an entertainment destination for millions from both sides of the Detroit River. Today the island falls under the jurisdiction of Amherstburg, a province of Ontario. And while the park is Canadian—the island has officially belonged to Canada since 1867—Detroiters claim it as their own in their collective memories. Had it not been for the amusement park, this island could have become as obscure as many others that dot the Detroit River, their names known only to avid boaters. But this island is different. It has become an extension of Detroit, even more so than the suburbs that eventually have grown out around the city.

Detroiters are perhaps more stubborn than surprised to acknowledge that the island does not legally belong to them. But Bob-Lo is Detroit—as Detroit as Vernor's soda and Stroh's ice cream. For it was primarily Detroiters who flocked to this playground in the channel, from its beginnings as a picnic area and resort for the Motor City elite to its last days as an amusement park for the minions.

To this day, Detroiters remain nostalgic about the park, and have kept its memories alive. The Internet has cemented its place in history as well. There are numerous Bob-Lo Island-related Web sites—for roller coaster buffs who know every ride that was ever at the park, to people reminiscing about riding the boats, to people collecting and selling Bob-Lo memorabilia on eBay.

Most folks remember Bob-Lo as an idyllic playground, but the island has a torrid history of disputed ownership, business deals gone bad and underhanded treatment of its residents. Like her two-story cottage that has stood for 46 years, Dorothy Tresness has persevered, continuing to live alone on the island and determined to hold onto her slice of paradise. For the woman who knows Bob-Lo most intimately, the story is more than just an island amusement park of days gone by. Most visitors would come for a day once or perhaps a few times over the summer. Dorothy came every day for nearly 50 summers and, even now at age 84, continues to spend three or four months there each year. Her memories include bringing over every single piece of the cottage by boat, from bricks for the fireplace to bags of cement for the foundation to rugs for the floors. She has been a fixture on her property for families and friends to visit, and a fixture in local Canadian governmental affairs as a frequent visitor at township meetings keeping abreast of developments concerning the island. Her rights to the land and the water are one of the few things that have withstood the tides of change on Bob-Lo Island.

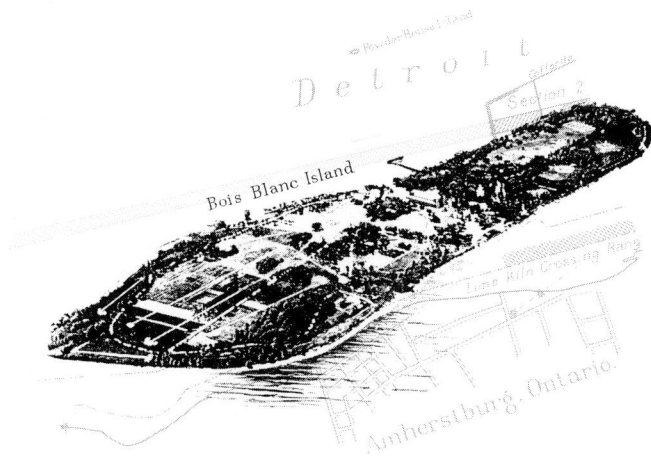

THE *I*SLAND IN THE MIDDLE

*"Mention the word Bob-Lo to anyone from the area in the last
four generations and it is sure to conjure up memories, good memories.
And that is the key to the island's existence. It represents the good times,
and everyone wants to be nostalgic for good times. ... As long as there
are memories, there will be a Bob-Lo Island." —Dorothy Tresness*

Bob-Lo Island is 272 oval-shaped acres situated in the center of the Detroit
River, actually a tad closer to Canada than to the U.S. It is longer than it is
wide, three miles long and only a half-mile wide, and sits at the mouth of the
Detroit River before it empties into Lake Erie—one of the five Great Lakes
whose waters touch Michigan, Ohio, Canada, Pennsylvania and New York.

The island's location has always proved to be significant; explorers navigating the Great Lakes would have to pass it, and military groups would use
it as a stopping point before entering the United States through Detroit. Its
first inhabitants were Wyandot Indians, sent from Detroit by the French mis-
sionaries who had founded the city in 1701. The French settlers referred to
the Wyandot tribe as Hurons, which meant Boars Head in reference to their
spiky Mohawks.

THE WHITE WOODS

Every Detroit schoolchild learns the story of the founding of the city by Frenchman Antoine de la Mothe Cadillac. They learn that he treasured this spot on the river as a gateway for transportation to the north and the south, and saw great promise in the abundance of natural inhabitants, the animals that would further the fur trade industry.

Where many history books stop short is in telling that successors of Cadillac established a mission on the banks of the river. They called it "Bois Blanc," French words for "White Woods" in reference to the abundance of white poplar trees. It was here on the sandy shores of the Detroit River that the French "kept" their Indians, continuing to convert them to Christianity and taking advantage of their prowess in living off the land, specifically in trapping for the burgeoning fur trading industry. Years later when the Indians threatened to flee to Ohio where English traders were settling, a treaty was drawn to actually give the tribes islands in Lake Erie and the Detroit River. Under this arrangement, the Wyandots—under French auspices—inhabited the outlying lands for many years. The largest of these islands took the name of the mission, and became known as Bois Blanc. The Indians' own name for it was "Etiowiteendannenti," meaning "a peopled island of white woods guarding an entrance." The Americans would come to call it "Bob-Lo."

As the French presence grew along the American side of the river, other Indian tribes were also displaced to the island. The tribes were led by various chiefs, some of whom went on to make their own mark in history. The more famous names staying at Bob-Lo included Black Hawk, a prominent player in the War of 1812; Shabbona, who went on to befriend the white man in settlement issues in Indiana and Illinois; Sagonash or "Captain Billy," the Indian son of Col. William Caldwell, one of the early settlers of Chicago; and Miera, or "Walk-In-The-Water," whose name was given to the first steamboat to cross the Great Lakes. From 1818 to 1821, the 135-foot long *Walk-In-The-Water* would carry up to 100 passengers at a time on a route from Buffalo to Detroit. Her historic service came to an abrupt end on November 21, 1821, when a storm drove her ashore near Buffalo. The waves pounded relentlessly until she broke apart.

SHIFTING BOUNDARIES

Many Detroiters and Michiganders are surprised to learn that today the island is Canadian, not U.S. territory. The detailed history of the boundary containing the island is complicated and has been disputed a number of times, beginning when the British overtook it from the French in 1796, and again at the end of the War of 1812 when Americans protested the British occupation. The very first engagement of that war actually occurred at Bob-Lo Island. An American ship, called the *Cuyahoga*, was proceeding to Detroit carrying the U.S. Army band, wives and children of officers, mail and supplies when canoes of Indians and British soldiers just off the coast of Bois Blanc overtook it. The Americans aboard did not even know that war had been declared.

Following this relatively easy capture, the British and Indians went on down the river to capture the fort at Detroit. Riding alongside with the British was a very capable Shawnee Indian chief named Tecumseh. The Wyandots and the Shawnee had a very special friendship, and Tecumseh used Bois Blanc as his headquarters, staying there frequently while gathering troops to help the British cause.

At the end of the war, the Treaty of Ghent attempted to determine some international boundaries between Canada and the U.S., including Bois Blanc. Documents decreed: "The boundary line is to run through the middle of Lake Erie until it arrives at the water communication between that Lake and Lake Huron, thence along the middle of the said water communication." The Americans interpreted "water communication" as the navigable channel—or the one usually used by ships—while the British claimed it meant the exact geographic middle of the channel. Since the west or U.S. side of the river only had a depth of a few feet and could not be considered navigable, the Americans claimed the boundary would fall between the island and the Canadian shore, putting it in American waters.

In 1818 a Boundary Commission was drawn in to settle the dispute. The Commission negotiated that the island would remain under British sovereignty. In exchange, the British gave up occupancy of Drummond Island farther to the north. The pilot of the boat used by the commissioners to map out the boundaries was James Hackett, a native of Scotland whose family had settled in Amherstburg. In later years, Hackett and his descendants would figure extensively in the story of Bob-Lo Island as keepers of the lighthouse.

Suddenly, as a result of the Boundary Commission's decision, the island loved by Detroiters became foreign territory. But over time, the 18 miles of waterway separating the two countries would not hinder millions of trips from Detroit to the beautiful island. In fact, the trip over water became part of the Bob-Lo experience.

Like Detroit, the British fort at Amherstburg—Fort Malden—sat right at water's edge. There was no protection, save for Bob-Lo Island sitting out a couple of hundred feet. In 1839, taking advantage of Bob-Lo's position as a defensive barrier, the British built three identical blockhouses along the length of the island. They were two-story structures made of logs, the top story projecting out over the bottom. Holes were made in the upper story large enough to shoot a cannon. This style of blockhouse was prevalent throughout the United States in the pioneer days.

When Fort Malden was disbanded in 1852, the blockhouses were no longer needed. The north blockhouse met its end in 1867 when it was torn down and used to make a bonfire to celebrate Confederation Day, which marked the unionizing of the four provinces of Ontario, Quebec, Nova Scotia and New Brunswick to form the country of Canada. The middle blockhouse became part of a dwelling that was in use until the 1920s. The south blockhouse withstood the island's transformation into an amusement park. In the 1930s it was a tourist attraction. Many years later it was converted to a souvenir stand. It and the lighthouse constructed in 1837 would become the longest-standing historical symbols on the island.

LIGHTING UP THE WATER

The lighthouse was erected on the island's southernmost point to help boats navigate through the narrow channel on the Canadian side. Constructed of limestone brought over from England, the lighthouse towered 56 feet above the water and guided traffic from Lake Erie through the channel to Amherstburg and Windsor. Its beacon—a 10-lamp oil-burner, with eight of the lamps facing toward the lake and two upstream—could be seen for 18 miles.

The lighthouse operated from 1837 until the autumn of 1954 when vandals broke in and set a fire that destroyed the lamp house.

One Amherstburg family kept those lights operating for nearly 100 of those years, a result of a rather unconventional appointment.

While in town to negotiate further treaties with the Indians, Sir Francis Bond-Head, the lieutenant governor of Upper Canada, had taken a fancy to a dog belonging to Amherstburg resident James Hackett. When he inquired if the dog, a massive Newfoundland, would be for sale, Mrs. Hackett retorted that she could not possibly sell the dog; however if her husband were appointed light keeper she would give Bond-Head the pet. Thus began a family legacy—as long as the lighthouse was manually operated, it was operated by a Hackett. When James Hackett died, his son, Andrew Hackett, took over. Upon his death in 1901, his widow was appointed, and she in turn was succeeded by her youngest son, Charles Hackett.

The only evidence of any military trouble at the lighthouse occurred within the first year of operation. In January 1838, the Patriots who had come down the river from Detroit to attack Fort Malden captured the island. The Patriots used the island as an outpost while their schooner *Ann* circled the island, firing cannons upon Amherstburg. Hackett and his family managed to escape from the island and took refuge in the back woods of Amherstburg. When they returned, they found that the soldiers had feasted on their livestock, but the lighthouse remained unharmed.

Today, tree limbs climb past the halfway point of the tower's height, and with no lamp, it looks like a strange monolithic symbol, oddly perched on the most narrow point of the island. In 1961 it became a national historic site of Canada and continues to be a landmark for boaters, and a point of interest for lighthouse devotees.

PRIVATE PROPERTY

The island's history as a privately owned parcel begins in 1860 when Colonel Arthur Rankin purchased 225 acres from the Canadian government for the sum of $40. Once the island was not needed for military operations, the government had used it as a political carrot. Originally it had been promised to John McLeod, a local merchant and Parliament member, in exchange for his vote on a legal issue. When he did not follow through with the vote, Rankin, his political opponent, was given the opportunity to pay the nominal taxes for the title to most of the island. Of the remaining acreage, 14 acres were given in a life lease to lighthouse keeper James Hackett. The government reserved the rest.

In 1869, the title for the 225 acres was transferred to the Colonel's son, Arthur McKee Rankin and his wife Kitty Blanchard. Arthur and Kitty were stage actors who became stars of American theatre. Arthur, much to his parents chagrin, had run away from school at the age of 16 and fled the country to join the theater in New York. By the age of 21 he was a leading man at the Arch Street Theatre in Philadelphia and had a flourishing career.

After marrying Kitty, the duo formed one of Canada's first theatrical touring companies and produced Western dramas, which Arthur wrote and starred in. He built theaters in San Francisco and New York City and even toured Australia and South Africa. He is affectionately remembered as McKee Rankin, the first great Canadian actor, even though he made his fortune by touring in the U.S.

In between weeks-long tours, Rankin would return home to Bois Blanc where he transformed his father's farm into a proper estate, a showpiece for the many friends the couple entertained. The farm was stocked with purebred Shetland ponies, cattle and deer. A small steam-powered yacht, the *Kitty B.*, would ferry their guests to and from Amherstburg. They enjoyed many years living the high life on the island, but a series of bad investments caused Arthur to eventually turn the title over to his wife to avoid creditors. He left the island and his wife and returned to the United States, where he continued to work in the theater. By the time of his death in 1914, he had become an alcoholic washed-up actor and died penniless. Soon after his departure from Bob-Lo, Miss Kitty had mortgaged the estate to local capitalist Napoleon Coste for $13,000 U.S. Upon acquiring the title sometime in the 1880s, Coste promptly sold the land to two Detroit businessmen, Colonel John Atkinson and James Randall—apparently at quite a hefty profit as the recorded price appearing in various historical documents ranges from $40,000 to $100,000.

With plans to develop the island, in March 1888, James Randall announced he would build a 100-room hotel on the island. A few months later, the *Amherstburg Echo* newspaper reported that Randall had actually built a system of streets on the island. This was a pre-cursor to the legacy Randall would leave behind as the creator of the boulevard system in Detroit.

Atkinson and Randall were close friends when they began planning development of Bob-Lo Island, but the friendship quickly went sour. In a major quarrel they divided the property, but Randall had already started constructing a home that fell smack dab in the middle of the new property line. Randall offered to buy the additional property from Atkinson, but Atkinson refused and demanded that his former friend tear down the house. Randall did not comply, only to come to the island one morning to find a pile of rubble where the framework once stood. In the middle of night a gang of men had arrived by boat presumably to do Atkinson's dirty work, creating what became known in Amherstburg as "Randall's Wreck."

Randall's troubles continued when his son, Tom Davey Randall, disappeared one night while duck hunting. Beside himself with grief, James Randall hired spiritualists and mediums to hold séances on the island during the winter in hopes of finding his son. The body was found on the banks in the spring when the ice melted. It is said that after that spring, Randall lost all passion for the island. Atkinson, meanwhile, built a marvelous structure—a gaming house for his sons—that would become an icon on the island. The house was round, with six bedrooms around the outside surrounding an open-room living space in the middle. A screened-in porch lined the entire perimeter. The house was known as the Owl Cottage, and in 1955, Dorothy Tresness and her husband Orin would rent it for five years and become its neighbors for another 40 when they built their own place next door. For nearly 100 years the Owl Cottage was a landmark on the island, its distinguished shape appearing in the background of generations of family photos. In addition to building the historic Owl Cottage, it was also Colonel Atkinson who gave the island its start as an entertainment destination.

RECREATION ISLE

In 1897 the Detroit Belle Isle and Windsor Ferry Company became interested in the island as a potential excursion spot. On December 15, 1897, it signed a lease for 25 acres near the middle of the island from the colonel. The lease was for 15 years with an option to buy, with a total purchase price of $250,000. In 1885 the company had begun day trips to Belle Isle, a smaller island closer to the city of Detroit, and had decided to expand its business with runs to other locales. To make Bois Blanc a destination hot spot, the company announced in *The Detroit News* that it would "transform the island into a pleasure ground for Detroit excursions by building a casino, observatory, pavilions, bath and boat houses, baseball diamonds, bicycle track and lawn tennis courts." The article continued, "The land is high, the surroundings picturesque and the air free from all city impurities."

The first excursion from Detroit was made on June 18, 1898 aboard a ferry named *Promise.* Excited passengers paid 35 cents for the round trip. By the early 1900s there were five boats in service, *Promise, Pleasure, Sappho, Britannia* and *Garland,* which was the first Great Lakes ship to have electric lights.

Getting to the island from the Canadian side was always a much shorter trip—just 875 feet across the water. Therefore, smaller boats could be used, as they could make more round trips. Ferry service from Canada began in 1902 when the Amherstburg dock was built. With access now from two countries, there was some concern about the island in the middle.

BORDER CROSSINGS

Early on, it was suspected that illegal immigrants desiring to enter the United States would do so via Bob-Lo Island. At one time in the mid-1920s, the Amherstburg ferry service was suspended until an arrangement was worked out that a U.S. Immigration inspector would be stationed at the Amherstburg dock. The customs issue continued for decades, with people's naiveté landing them in legal situations as they tried to enter or leave Canada without the proper identification.

In the early part of World War I, a German terrorist who was wanted for blowing up a Windsor factory that was producing war goods was hiding in Detroit. The Canadian Mounted Police, working undercover, befriended the man and persuaded him to take a trip to Bob-Lo Island. Of course, the minute he stepped onto Canadian territory, they had their man. Four decades later, a story appeared in a Detroit newspaper about a Wayne State University medical student from Kenya who arrived at Bob-Lo with an expired student visa. His trip to the amusement park ultimately cost him his chance at finishing his education as he was extradited to his native country.

BOATS, BANDS AND BOOZE

As much as it represented a security threat, there is no denying the effect the amusement park had on the Amherstburg economy. In the early 1900s, the park was promoted as a family and church picnic ground. The park was open only during daylight hours and liquor was strictly forbidden. It was not uncommon for some thrill-seeking adults to become bored with the outings and convince the ferry captain to take them for a quick jaunt to the Canadian side. They indulged and overindulged themselves at the hotels and inns along the riverfront. During the Prohibition era from 1920-1933, the port served as headquarters for many rumrunning gangs attempting to get their product to the Detroit shores. The American appetite for booze was so great that a Seagram's V.O. distillery was built just north of Amherstburg in 1928. The distillery had its own small airport, so the liquid gold could be delivered by air as well as water. The original plant burned down in 1950, but was rebuilt and is still in operation today.

From the beginning, the park's popularity was so great that the small boats could not keep up. In 1902 the *Columbia*, which held 2,500 passengers, set sail. The boat would continue its Bob-Lo run for 89 years.

The trip from the Detroit docks was over an hour, and from the beginning efforts were made to make the trip aboard the huge steamer *Columbia* entertaining. Live bands, dancing and dining were part of the admission price. The first notes of "Anchors Aweigh" bleating from the instruments of the Harry Zickel Orchestra and a single long blast of the horn signaled to all it was time to get under way. At a time when most people had never been on a boat this size before, just walking onto the *Columbia* was an experience. The whitewashed ship, designed by legendary naval architect Frank Kirby, was 185 feet long and four stories tall. Its rails and floors were made of mahogany, encasing the belly, which held a 2,000-horsepower steam engine.

As the orchestra played, guests were free to dance or walk about the ship, or go up to the beer garden on a separate deck. On the return trip children would often fall asleep across the seats while rows and rows of red life jackets hung from the rafters above their heads, cushioning their dreams of the day they just had. Moms and dads would find a few minutes to sit peacefully and look at the stars, and perhaps imagine for a moment they were alone on a romantic cruise ship like the *Queen Mary*, the likes of which they had only seen in the papers. Young men and women on dates would imagine they were alone too, stealing a tentative first kiss under the stars.

In 1910 the *Ste. Claire,* also designed by Kirby, joined the *Columbia.* From outward appearances, the ships looked essentially the same. Both had engines of 2,000 horsepower and could hold 2,500 passengers. Ironically, although she stretched 30 feet longer and was newer, the *Ste. Claire* suffered a bad start. During the inaugural setting in the water, she listed to one side and from then on was always considered not quite as good as the original *Columbia,* much like an ugly stepsister. Oftentimes, passengers waiting at the Detroit dock would see the *Ste. Claire* approaching, recognizable by four round port-holes on its bow and its flying bridge atop, and would refuse to board, insisting on waiting for the *Columbia.* In the end, however, it was the *Ste. Claire* that got the glass slipper—it was purchased by a private investor and lovingly restored while the *Columbia* was left to sit in limbo, literally rotting away in the Detroit River.

The two big ships were owned by the Detroit Belle Isle and Windsor Ferry Company and were docked in downtown Detroit. They became known not by their proper names, but simply as the "Bob-Lo Boats." With two boats running at full capacity all summer long, beginning in 1911 more trips from Detroit were added. The ships had also become an attraction in of them-selves—cruises that did not even stop at the park were also popular. A moon-light cruise and a Sunday lake cruise were offered. By 1912 the night cruise was offered Tuesday through Saturday, and trips to the island were 9 a.m., 1:30 and 3:30 p.m. with return trips at 2 p.m. and 8 p.m. Prices were 35 cents for adults and 25 cents for children.

ADDED AMUSEMENTS

The owners began to use the profits from the boats to improve the park. A new dance hall, the largest in North America, was erected. Made of steel and stone, its entrance looked like a cathedral with a wall of glass bordered by two stone pillars. Dancers were charged 5 cents a couple and there were rules. No "wild" dancing was allowed, no turkey trot, bunny hug or bear dances. The boats continued shuttling passengers May through Labor Day and the park continued to expand. By 1923 prices had gone up to 60 cents for the boat ride and there were more attractions on the island. A new cafeteria that seated 250 people was built to overlook the water, and a golf course shortly followed. The main attraction continued to be picnicking, dancing and outdoor sports. Tennis courts, a boathouse, a children's playhouse and a women's Rest House were built around the beach area.

The first true ride was a carousel that featured an organ and hand-carved horses from Italy. It was brought over by boat in pieces, and when constructed it was deemed to be the largest carousel in the world. As the bicycle was becoming extremely popular, a one-third-mile track with banked walls was built for racing and bikes were available for rent.

Long-distance swimming competitions became all the rage, and swimmers set records from Detroit to the island. For years, Detroit teenagers Dorothy deCaussin and Ida Mutnick held the record of 8 hours, 42 minutes. The girls were 17 and 15 years old, respectively.

By now, Detroiters had given up pronouncing the French island name Bois Blanc. Crude attempts usually sounded like "boys blank" and some even took to shortening it to "Blah Blah." The island was christened Bob-Lo; the Americanized name would stick for the rest of time. The name was hyphenated until the 1980s when park owners copyrighted it as one word—Boblo. The most recent owner reverted to using Bois Blanc to capture the historical significance of the island and its international flavor. But for its longest private resident, it has another name.

Mrs. Orin

T

LEASE

Tresness

𝒜 NEST OF TREES

*"The men would bring their wives and kids, and others would
often drop in either with their own boats or by hitching a ride on the
Bob-Lo Boats. The kids would swim and swing from tire swings out over
the river. Life jackets were kept on an iron pole at the top of the hill and
no kid went over the hill down to the water without one. Other rules were
no child went to the water alone, and no shrieking in the water.
No crying wolf, yelling 'Help! Help!' Not minding the rules
would result in two days restriction in the house.*

*"They would come in on Friday night and party a little bit, work
all day Saturday and then party a lot on Saturday night. We would
go to Canada and get the Canadian beer, oh how we loved that
Canadian beer. And on Sundays everyone would recuperate.*

*"The times truly were good. There were no incidents or accidents,
and wouldn't be for nearly 40 years." —Dorothy Tresness*

Dorothy began visiting the island when she was just 6 or 7 years old, or maybe even younger. She remembers going over on the Bob-Lo Boat with her mother from Detroit, where her family had settled. It was the early 1920s, and at that time the park consisted of picnic areas, a beach, swings and slides and a merry-go-round that had to be pushed. Dorothy's mama would always take her to visit the families that owned property on the east side of the island. Four lots were owned by three brothers and a sister: Hubert (Huey) Monahan, Thomas Keena Monahan, Leo Monahan and Annie Mae Monahan (Quinn). The fifth lot was owned by Fritz Hodges, who may have been related to the Monahans, possibly a half brother. The sixth lot was the Owl Cottage, owned by a family named Geoux. Together the six lots created a little resort community that the residents called "Tanglewood." The Monahans and Geouxs were upper-class citizens and enjoyed many of the trappings associated with a privileged lifestyle. Their properties on the island included tennis courts and sailboats. Leo Monahan was the deputy controller for the city of Detroit and very influential in Detroit politics. His frequent guests included auto tycoons Walter O. Briggs and William Fisher. The public picnic grounds covered a small part of the island; it had really become a retreat for Detroit's rich and famous.

Photos in scrapbooks kept by the Monahans, and now passed on to Dorothy, show the Detroit Women's Historical Club visiting in 1927, the women wearing skirts down to their ankles and furs in July. When the Detroit elite were there, "letting it all hang out," children were not allowed, so Dorothy would board the Bob-Lo Boat for the ride home at dusk, before the wild parties began.

"One famous story that got passed down was that Mrs. Fisher had lost a seven-carat diamond ring on Leo's tennis court," Dorothy recalls. "We spent quite a few summers looking for that ring, but it never did show up. I imagine Mrs. Fisher just got herself a replacement."

Raymond Geoux was a doctor in Detroit, with a practice in the David Whitney Building downtown. He and his wife Charlotte had a son, Warren, and a daughter, Louise. Louise was just a few years older than Dorothy and the two girls became as close as sisters. Young Dorothy grew up in this extended family, knowing the Monahans as Uncle Leo, Uncle Tom and Uncle

Huey, even though they were not related. In those days, children would not think of calling an adult Leo or Huey sooner than they would try to fly to Egypt, she quips. The men doted on Dorothy and her family, and the ties remained for life. Leo died in the 1950s. Tom Monahan eventually moved to Florida and continued to write to Dorothy until the 1970s, and occasionally came back to visit the island.

Louise Geoux, her childhood friend, would become the reason that Dorothy is still on the island today.

A NEW LIFE TOGETHER

In 1942, Dorothy married Orin Tresness, who was from Wisconsin. They met that year while Orin was serving as a military policeman at Fort Wayne, on the south side of Detroit. Dorothy had gone to a party at the Fort where Orin was on duty. They caught each other's eye, but agreed to be just friends, as both were engaged to other people at the time. More than 60 years later Dorothy can still recall the name of her Orin's fiancée—Cecilia.

"He was 23 and I was 22. That summer he took me home to meet his mother and she liked me. She told him I would make a good wife. So when we got back from Wisconsin he said, 'Why don't we get married?' " They both "took care of a little business" with their former fiancées.

Dorothy's beau had been shipped out to San Luis Obispo with the Air Force and she mailed his ring to him. He went on to marry a gal in England and years later settled in Florida, near where Dorothy and Orin bought a winter vacation home. He saw Dorothy's picture in a newspaper in 1977 when she was named Professional Secretaries International Secretary of the Year and the two had a reunion dinner. It was an amicable meeting and a happy ending, and Dorothy likes happy endings.

When they were first married, Dorothy began to bring Orin to visit Louise and the others on the island on a regular basis. Orin—who became known exclusively as "Tress"—had grown up on lakes in Wisconsin and loved being around water. It didn't take long before the island caught his heart as well and the young couple would soon have a place to call their own in Tanglewood.

OWL COTTAGE

Owl Cottage had stayed in the Geoux family, passed on to Louise and her brother Warren. Warren sold his half to Louise, who married and settled in Connecticut. She refused to sell the Canadian cottage, telling others that if war ever broke out in the U.S. she would have a safe place to retreat to.

One day in early fall of 1955, while fishing for walleye, Tress pulled his boat up onto the banks of the island to have lunch. It was there that he found a dilapidated sign lying in waist-high weeds in front of the Owl Cottage that said "For Rent." The phone number on the sign led back to Louise Geoux, now named Louise Lovekin. After sitting empty for 20 years, the cottage was rented to Dorothy and Orin. The couple, who now had two very young daughters, were about to make a sizeable investment in money and hard work.

The lease was signed on November 8, 1955. Conditions of the agreement included that the couple would:

A) "Recondition the lawn by cleaning out all weeds and obnoxious growths and raising a grass lawn."

B) "On or before November 1, 1956, put in new footings and level the house, put in new joists and new roof boards where necessary, replace any missing window glass, repair door casings so all doors will close tightly, put on new door locks, replace screens on porch where necessary and put on two new screen doors."

C) "On or before November 1, 1957, to paint inside of house throughout, and also paint porch floor."

D) "On or before November 1, 1958, to paint exterior of house and out-buildings."

When Dorothy and Tress opened the door for the first time they were met with a scene fit for a horror movie. The cobwebs hung heavy and the straw matting covering the floor had grown into a wall-to-wall carpet of living moss full of wiggling, writhing creatures. Undaunted, the Tresnesses brought 30 gallons of paint and an air compressor over by boat to get the old cottage up to snuff.

Owl Cottage had never been renovated and its original design was intact. It was designed for entertaining, built as a round house, with six bedrooms along the outside perimeter. Five of the bedrooms had two doors each, one from the living room and one to the screened-in porch. The sixth bedroom, closest to the kitchen, had been the maid's quarters and was the only one that did not have a door to the porch. At the center of the structure was a large 25-foot square dining room, with a table that had six leaves in it and could seat 20 people. The room was used for dancing, card games, whatever the crowd wanted to do. And there was always a crowd.

Not a weekend went by without all rooms at the Owl Cottage being filled. Dorothy recalls that many uncles, brothers and friends would come, at first to help clear the lot of sumac and overgrown trees, and later to help with construction of a new house next door. Sisters and nieces who would come to watch would be put to work as well. Once or twice a summer a yearling would be bought for a pig roast. The men would stay up all night turning the pig on the spit above a 6-foot square barbecue pit.

BUILDING TRESNESS

Early into the lease of the Owl Cottage, the lot next door became available as owner Fritz Hodges—who Dorothy knew as Uncle Fritz—died.

Fritz, a former steamboat captain on the Great Lakes, had suffered a stroke leaving one side of his body disabled. While burning leaves one day, he caught himself on fire. His pet, a large St. Bernard, kept any would-be rescuers at bay and Fritz burned to death, the scent of his prized peonies mingling with smoke and burning flesh wafting across the island. Fritz had no wife or children and the Tresnesses became keepers of his estate. They paid some debts associated with the property, and in return were given the 50-foot-wide lot. This was the lot where they chose to begin building their own cottage.

In 1956, the first year they had Uncle Fritz's property, the foundation was poured. In the second year the four walls and roof were completed. During construction, Annie Mae Quinn sold the Tresnesses her 50-foot lot next door. Dorothy remembers signing the papers at the hospital as Annie Mae was on her deathbed. The Tresnesses now owned two lots, leased the Owl Cottage and owned their home in Inkster on the Detroit side of the river. They referred to their spread on the island as "Tresness," the family name that also means "nest of trees" in Scandinavian.

Construction took nearly five years. It was a slow process, as everything had to be brought over by boat from the Detroit side and work could not be done during the harsh winters. The couple was also determined to pay for everything as they went along. The foundation was dug with hand shovels; bricks for the fireplace were brought over one load at a time. "To this day I can't look at a brick as a brick, it is always three-and-a-half pounds," Dorothy says. Lumber and roofing materials were brought over by barge by the Browning family, then owners of the park and very good friends to Dorothy and Tress. The Browning barge would also take the Tresnesses to church across the river every Sunday.

The building of the cottage became a true community event on both sides of the river. In Grosse Ile, the community downriver of Detroit where Tress docked his boat, the regulars at the local bar held a farewell toast to the Tresnesses when they were about to bring over the last 100-bag load of concrete and a cement mixer. Dorothy recalls, "They said, now we have seen it all—we have watched you take over 30 gallons of paint, a stove, and bottled gas, but a cement mixer? Here's to Tress, whose resting place is sure to be at the bottom of the river after tomorrow."

The cement mixer, and Tress, arrived at the island in perfect condition. The cement began pouring and the cottage was on the rise. Work continued from May through September each year, and in August 1960, they moved in.

Once settled in, they relinquished the lease on the Owl Cottage to another family, the Macys. At home in the Detroit suburb of Inkster, the Tresnesses lived next door to a relative of the Macys and knew the family quite well. The Macys and their five children would become part of the island community for the next 40 years. The lone Macy daughter became like a sister to Dorothy's two girls.

INDEPENDENT SPIRIT

Just when the Tresnesses could begin to enjoy the fruits of their labor at their island retreat, serious health issues began plaguing Tress.

"Tress was working for Darrin & Armstrong, a construction company," Dorothy says. "We were scheduled to go to Hawaii, Tress was going to be in charge of building the Sheraton Hotel when we found out he had to have surgery." The surgery turned out to be a double aorta transplant—new territory in 1960. Tress would be the third person in the United States to undergo the procedure. He came through, but would never completely regain his health. He then became what Dorothy calls a "government guinea pig" and participated in a series of drug trials, one which turned out to be testing coumadin. Although she was doubtful, Dorothy admits the rat poison worked and his heart began doing a better job pumping blood, but the rest of his body could not keep the blood circulating.

In a second life-threatening operation, arteries from the groin down were replaced with synthetic ones. He was only the sixth person to ever have had this done, and the first to become a living success. On the 20th anniversary of the operation, he made medical history and his case appeared in *The Journal of the American Medical Association*.

After the two operations, Tress could no longer work. Dorothy went to work as an executive secretary. In the summers, her mother would take the Bob-Lo Boat over on Sunday night and stay for the week to help with the two girls. Dorothy would drive her own boat back and forth from work. Come Friday, mother would get on the Bob-Lo Boat back to Detroit to spend her weekend socializing with her friends before returning on Sunday. Dorothy had learned to drive a boat early in their marriage, at her insistence so she would never have to depend on Tress or others for transportation.

"I didn't know a bow from a stern when we first got a boat, but I was gonna learn. My husband didn't want to take me out and I said, 'well, too bad,' and I took that cruiser out into Lake Erie with my two little girls and I had never drove a boat in my life. The trouble was when you put it into reverse the handle on the motor came up and you had go to the back of the boat and push it down. Well, I had put it in reverse and was doing circles in the middle of Pointe Mouille. I didn't know what to do and then I finally remembered seeing Tress go to the back of the boat and doing something. I figured it out all right—I've always been a pretty self-sufficient lady."

The self-sufficient lady earned her sea legs rather quickly. Within a few years, the couple had his and her boats. Dorothy not only learned how to drive a boat, she did her own maintenance and in later years even bought her own boats. Living on an island, life is dependent on and often dictated by the water. With a liquid highway separating her from her family, friends, stores and other necessities, Dorothy learned to revere the water and respect the weather. For 40 years she logged the daily weather, noting the temperatures and the wind direction. Her niece once gave her a windsock in the shape of a penguin. Dorothy promptly named him Petey and spoke of him like one does of a pet. Her journals include page after page of weather-related notations, usually written according to Petey.

"8 a.m. skies clear, River very calm, only the little ripples caused by the current. Petey is almost inactive but the radio said we have winds from the NE at 10 mph. All I see out there are trollers. They predict clear skies with a high of 84 degrees and a low of 60 degrees tonight with possible showers … Petey has started dancing a little and agrees the winds are from the NE, but still are not 10 mph here.

"9 a.m.- Petey is at 210 degrees, so the winds have arrived."

Many summer entries included the sudden appearance of a storm that would change Dorothy's plans for going stateside that day. On rainy days, her journal entries would often be lengthier, a time to reminisce about living on the island and sometimes being held captive by the weather.

"I remember once before when I was taking the girls to church on Grosse Ile when the fog rolled in and we couldn't see the bow light. We just dropped anchor and waited it out. That's a scary feeling sitting in a boat, surrounded by fog, seeing and hearing nothing. If you do hear anything the sounds are distorted and magnified and it's next to impossible to identify the direction from which they are coming."

Boats would always be an integral part of life at Tresness, and would continue to be the way for the masses to experience Bob-Lo Island.

W. B. Browning,
Vice President and
General Manager

THE *B*USINESS OF AMUSEMENT

*"As a child, I went over on the Bob-Lo Boat with my momma to go to
the park. There were merry-go-rounds that you had to push, swings and
slides for the children. It was strictly a picnic park when the Detroit Windsor
Ferry Company owned it. Now when the Brownings bought it, they were
forward thinking and built the amusement park using electricity. They
really made Bob-Lo the park we all remember." —Dorothy Tresness*

In 1949 the Browning family bought the park and boats from the Bob-Lo
Excursion Company (formerly the Detroit Windsor Ferry Company). On
opening day, May 28, 1949, there were seven sailings a day from the Detroit
docks. Round-trip rates were $1.30 for adults, 65 cents for children. From
1949 to 1959, the Brownings increased the number of rides at the park from
six to 27. Improvements included a new miniature train, a fun house and
a petting zoo. Roller coasters—including the Super-Satellite Jet and Wilde
Maus (Wild Mouse)—were imported from Germany and started attracting
enthusiasts from all over the country.

In 1957, the Brownings experimented with chartering a ship named the
Canadiana to bring customers from Toledo. The three-hour route had run

for only two months when the ship ran into a bridge. Fortunately, none of the 2,740 passengers on board were injured, but the incident put an end to that marketing strategy.

A CHILD'S PARADISE

While the park was growing, it did not impede on Tresness, nor the family's sense of seclusion. The ninth hole of the golf course at the back of their property was the closest sign of the park. The golf course had been there since 1926, and as young teens, Louise and Dorothy would peddle cool drinks to the golfers. Most people visiting the island had no clue that people lived behind the amusement park, and for Dorothy's two daughters and the Macy kids, sneaking into the back of the park became a closely held family secret. Years later, they would tell their own kids of the adventures—navigating through overgrown sumac and woods, then past a 6-foot chain link fence that had been erected by Canadian Immigration. The fence was locked, but the government gave Dorothy the key. Her girls would simply unlock the gate and ride on the rides. They would give their spending money to vendors for pops and ice cream, and their adolescent hearts to good-looking boys for the day. The girls befriended many park workers and were allowed to roller skate for free. After skating for hours they would approach the train station where Mrs. Hamilton, a short woman from Amherstburg with tightly permed white hair, would let them pass right through the line and onto the train without a ticket. The older lady seemed to look out for the girls as she always told the engineer to "take her slow down by the pony barn and let the ladies jump off."

And even when the park closed after Labor Day, the island held other forms of amusement. The United States and Canada would often hold mock wars or military exercises behind the park. The top of the cottage roof provided an excellent vantage point, until the girls' giggling gave them away. They would watch the men shoot blanks at each other, and the medics rush in and administer care. And then came the best part—collecting the ration cans left behind and having a feast by the campfire.

For the Tresness and Macy brood, growing up on the island was a child's paradise. Days were spent playing kick the can, digging up night crawlers, skunk hunting at the dump and learning to water ski (on one ski—the parents could not afford two). When the marina was completed, park owner Red Browning no longer wanted the kids water skiing around that area. One day, climbing onto the bow of his boat with his back to the freighter channel he began yelling at the youngsters to go the other way. Much to the kids' amusement, just as he was waving his fists in the air a huge freighter came behind him and its wake knocked him overboard. One of the Macy boys jumped in and saved him; not another word was ever said about water skiing.

THE BROWNING ERA

Lorenzo "Red" Browning was the driving force behind the park for 30 years, and a true ally of the family that owned the little white cottage. In addition to being a partner on the island, he also owned part of the T.H. Browning Steamship Co., and it was his barge that brought over the lumber and larger items used in the construction of Tresness.

Browning wore many hats as far as careers go, but the title of park owner was by far his favorite. In addition to the steamship line, Browning also dabbled in politics. He served on the city council of the affluent Detroit suburb of Grosse Pointe from 1971 to 1983, was then elected mayor of the city, holding office until he died. He had never lost his sentimental attachment to the island. Just one year prior to his death, Browning attended the auction when the island was for sale. He admitted that although he didn't have the $3.7 million to buy it back, he wanted to see what was going to happen to it. He was 79 years old when he died in his sleep in his West Palm Beach, Florida, home.

The Brownings had inherited a park that had suffered economically for a number of years. During the Great Depression in the early '30s, the park had actually closed for two seasons. It reopened in 1935, but attendance was relatively low throughout the rest of the decade. People were taking more vacations by automobile, and the building of the Ambassador Bridge and Detroit-Windsor car tunnel now provided a more convenient route from Michigan to Canada.

As World War II erupted, fewer Americans were venturing out to the island and Canadians were required to have a passport to visit, even though the island was Canadian. The boats crossed American waters and the Detroit officials were worried that people might travel to the island on a Canadian ferry and then hop aboard the Bob-Lo Boats to enter the U.S.

Other incidents had damaged the park's reputation. On June 9, 1941 thousands of Sunday School children were left stranded on the island at closing time. The churches, which had originally estimated attendance at 3,200, actually brought 10,000 children for the day. At closing time, the ferries had to make multiple trips to return the groups to Detroit, with the last boat pulling up early the next morning to a dock full of angry parents. A report in the *Windsor Daily Star* described the scene:

"Harrowing scenes were enacted on the dock this morning as frantic parents milled about helplessly, worried sick over the fate of their children. As always, wild rumours of boats sinking and other terrible tragedies flew about, and there was none to deny them. The last boat arrived about 3:30 a.m. and as the boat finally docked and the children arrived fright turned to anger. Seen through the eyes of children who did make it to school the following day, the night ride was a horrible nightmare in which they stood crushed and almost suffocated for hours before a late trip up the chilly river that left them chilled to the bone. Children reported people fainting, babies crying and many almost freezing to their death on deck of the boat when they could not fit into the cabin."

This incident, however exaggerated, was followed by a short labor strike by park employees who claimed unfair treatment. Shortly thereafter, a 14-year-old girl fell from the park dock and drowned. And, in 1945, six Detroit street gang members attacked a waiter on the Bob-Lo Boat and were arrested.

CLASSIC COASTERS AND THE CAPTAIN

When the Brownings began operating the park in 1949, they immediately began making improvements—adding to the dance hall, installing six adult rides, a skating rink and a miniature train. Many of the rides acquired during this time would stay in the park for decades, including the Caterpillar, Tumble Bug, Whip and Comet. On the ferries, Joe Short, a performer with Ringling Brothers circus, was hired to play the part of a clown and entertain passengers. Many Detroiters had their picture taken with the diminutive Short, who stood no taller than the average 8-year-old, wore a top hat and carried a scepter. He became "Captain Bob-Lo" and worked on the boats for 25 years. He was 99 years old when he retired in 1974. He is buried alongside Detroit dignitaries and celebrities in the city's famous Elmwood Cemetery.

In the 1960s the Brownings sunk the lake freighter *Queenston* so that its deck could be used as a walkway, replacing the wooden dock. Parkgoers would then proceed under a covered canopy several hundred feet long before reaching the park's entrance. A zoo was also added with 300 animals. In 1972 three baboons escaped and roamed about for a few hours, causing havoc and making the newspapers. By 1970, annual attendance approached 1,000,000. Throughout the next decade, the Brownings continued to add new rides, restaurants and free shows.

Later rides included the Swiss Toboggan, Meteor, a log flume ride, Rotor and Sky Streek, which would become the park's most infamous roller coaster. Made of steel, the Sky Streek replaced an earlier wooden coaster, Thunder Bolt, that had been in the park since the 1930s. Over the years the cars were painted different colors, but most of the track remained white. It was a classic out-and-back, oval-shaped track. The Japanese-made steel coaster is still running today at Selva Magical Park in Guadalajara, Mexico, where it is known as "El Cascabel" or Rattlesnake.

The Brownings had taken a park that in the 1940s could have closed because of wartime economics and a negative reputation and transformed it into a premier entertainment destination. By 1974, the Brownings again tried to reach out to a new market, acquiring a used tour boat, considerably smaller than the larger Bob-Lo Boats, and began running service from Wyandotte, a Detroit downriver suburb. The Wyandotte location was much closer to the island than the Detroit dock, and gave the residents more access to the island. The ship was officially named *City of Wyandotte*, but many riders referred to it as "The COW." Service from Wyandotte became a key in years to come as future owners struggled with the amusement park; the Brownings had set a precedent and the downriver residents had become an important customer base.

With all the continuous improvements made, it came as a shock when the Brownings put the park up for sale after the 1978 summer season. Attendance had gone down slightly, and speculation arose that the park had suffered from an increasingly unsafe reputation. Thirty visitors had been injured and one killed over 13 years.

The first recorded major accident under the Browning reign occurred in August 1965 when one person was killed instantly and eight others injured on the Tumble Bug when a car literally flew off of the tracks. An investigation cited faulty construction and a $750,000 lawsuit was filed. The ride was repaired and put back into service. The Brownings agreed to settle out of court for an undisclosed amount.

The island made the news again in the summer of 1973 when a girl was thrown from the Wild Mouse. She survived, as did two other people who were thrown from the Sky Streek. There were also accidents in 1974, including a park employee who stepped in front of a roller coaster and a mother and daughter who got the ride of their lives on the Scrambler. Both had significant injuries when the car they were riding in came loose in mid-air and went tumbling end over end down to the ground. In 1976 the new Galaxi roller coaster made a dubious debut. Two employees suffered injuries while testing the coaster and a short time later a girl was thrown from the ride due to improperly set brakes. That same summer, a maintenance man got pinned between two cars on the Galaxi when a ride operator did not stop a loaded car at the platform. He survived, but it was surely a ghastly ordeal for the people riding in the car that hit him. Two years later, the Sky Streek again had a problem, this time injuring 10 riders in a rollback accident. Also that same year (1978) two young girls were hurt when they fell from a cable car.

NEW OWNERS, NEW PROBLEMS

The Browning brothers wanted out of the seasonal business, and as their children had no desire to carry on the family tradition, the park was put up for sale. In early 1979 a consortium of owners operating as the Island of Bob-Lo Company purchased it for $5 million. The company immediately set about with its own vision, converting the roller rink into the Carousel Theatre that housed musical reviews—including a 1950s show with an Elvis act. A new Maxivision stand-up theatre was also touted in brochures for the park that year.

Despite pumping an additional $5 million into the park, on opening day 1980 only three people boarded the Bob-Lo Boat in Detroit. Attendance for the year plummeted from 500,000 in preceding years to less than 300,000. Bookings for company and school picnics, once the major source of income for the park, dwindled. The park was also feeling competition from bigger amusement parks—such as Cedar Point just two hours away in Ohio—that had come to dominate the industry.

Before even opening again in 1981, the Bob-Lo Company and its parent company, Cambridge Properties, filed for bankruptcy in both the U.S. and Canada. Court documents listed the amount owed to creditors as $5,395,000 (Canadian) with an additional $496,000 (Canadian) due to the government and payroll. Creditors appearing on the bankruptcy papers, which Dorothy secured a copy of, included $60,791 to the Joel Theatrical Rigging Company of Ontario and $30 to the Better Made Potato Chip Company in Detroit.

THE AAA ATTEMPT

The park struggled until The American Automobile Association of Michigan (AAA), an insurance and travel conglomerate, stepped in and purchased it for the 1983 season. For AAA, the purchase of the island was as much about politics as about profits. The Detroit-based company had chosen to relocate its headquarters, and thousands of jobs, outside of the city limits. Faced with political pressure, it made a commitment to maintain a presence in Detroit.

The commitment was fulfilled with a downtown warehouse for the amusement park that included boat storage. Within a few years, AAA poured $15 million into the park, changing the layout, improving the concession offerings and adding a corkscrew roller coaster called The Scream and the Sky Tower observation structure. The used Constram-built tower took riders up some 300 feet and revolved to give them a 360-degree view of the island and shores beyond. The tower is the last visage of the park that is still in existence today, the most recent owner left it as a landmark with plans to operate it on special occasions. AAA also added service from another down-river community, Gibraltar, with new Mississippi-style river boats. Its travel division heavily promoted the park as a destination for families as well as corporations, and attendance once again approached 1,000,000.

But the economies of scale apparently were not in balance to keep the seasonal park open, and AAA wanted out after just five seasons. The company had continually poured millions into the park's physical structures and marketing efforts but, like its predecessors, was unable to control some aspects of the public attraction. The crowning blow came when gangs of Detroit teenagers were creating problems in the park and on the boats. In its last days of ownership in the spring of 1988, AAA officials called in the Ontario police to oversee loading of the ferries at closing time. Gang skirmishes that began at the park continued on the water, with people actually thrown off the boat and falling into the dark river. At least 29 people were reported injured one night, including one boy who had to be pulled from the water and resuscitated. An off-duty police officer claimed he witnessed a girl being beaten and thrown over the side and a pregnant woman reportedly had a miscarriage as a result of the chaos. This gave the park a week's worth of bad press in what was labeled as "the Memorial Day incident."

Fortunately for AAA, they were able to duck out of this PR nightmare and avoid another season of trouble—the papers had already been signed and the island officially had a new owner. International Broadcasting Corp. had bought Bob-Lo for slightly more than $20 million.

CUTTING COSTS AND CAROUSELS

IBC owned the Ice Capades and the Harlem Globetrotters touring shows. This was its first foray in the amusement park business. Owner Thomas Scallen saw buying Bob-Lo as a natural extension of his products focused on family entertainment. He also thought the seasonal park would bring in revenue when the ice skating and basketball shows were on summer hiatus.

The first summer proved to be baptism by fire. The infamous gang incident had resulted in a 10 percent drop in business for the summer. Scallen responded by beefing up the island security presence, literally, with muscled bouncers who were very visible in the park. His pet name for the staff was the "Bob-Lonians." He also bought a speedboat and designated it as a paddy wagon to quickly remove troublemakers from the island.

The park now had three major roller coasters as its star attractions—Sky Streek, Screamer and Nightmare (an enclosed coaster now called the Mayan

Mindbender at Astroworld in Texas). Using his entertainment background, Scallen created more shows at Bob-Lo, including the International Pavilion Dance Show, Incredible Acrobats of China, the Ski Fever Water Show, magic shows and animal shows. Seven souvenir outlets sold T-shirts, mugs and ashtrays. There were two full-service restaurants and 15 snack bars. By all accounts, the park was still operating full fledge, but again there were signs of underlying economic stress.

In February 1990, IBC sold off pieces of the carousel that had been on the island since 1878. The carousel, which had been restored and returned to service in 1978 on its 100th anniversary, was a premier example of work by the most famous team in carousel design—Mangels-Illions.

William Mangels was a carousel designer who built the frames and mechanics for carousels. He commissioned Marcus Charles Illions, recognized as one of the greatest carousel artisans of all time, to carve the animals. When finished, Illions had carved 44 jumping horses, two goats, two deer and two chariots for the Bob-Lo carousel.

The goats were the only goats that Illions ever made. Attention to detail and the ability to capture motion were Illions' trademark. Through the position of the head and legs with manes that seemed to be whipping in the wind, he froze the essence of speed. He also used 22-karat gold leaf and jewels to decorate his magnificent animals.

The carousel was dismantled and brought across the Detroit River in pieces to a hotel in Dearborn, Michigan, where it was auctioned off. One deer went for $34,000. A horse fetched $21,500. When all pieces were sold, it netted $5 million for the park owners.

The next season IBC announced that in order to cut costs the park would be closed on Tuesdays. It was the first time in 93 years that the park would not be open seven days a week. The public would deem the next money saving move unforgivable—IBC took the Bob-Lo Boats out of service.

THE BOB-LO BOATS

For years, many friends and family used the Bob-Lo Boats to visit Tresness, and Dorothy would buy season passes for her teenaged grandchildren to get back and forth. The grand old steamers—the most recognizable symbol of the island for nearly 100 years—were gone as of 1991, but would not be forgotten. They have a history almost as complex as that of the amusement park. But while the fate of the amusement park has a definitive end, the fate of these cherished vessels would continue to drift.

IBC's announcement that the boats were no longer going to run from downtown Detroit and service on smaller boats would only be available from Gibraltar and Amherstburg was not taken lightly by the city. Former Mayor of Detroit Coleman Young went so far as to call the decision "racist." While speaking at a conference of southeastern Michigan mayors, Young referred to a memory he had of being denied admission on board a Bob-Lo Boat as a youngster in the 1930s, and stated that the current owners had reverted to a similar kind of behavior.

IBC maintained that the boats cost the company more than $2 million each season. That included maintenance and operating costs for a live-on crew of 31 people, some of whom had made a career on the boats. One of those people was Linwood Beattie, captain of the *Columbia* from 1944 to 1981. When ferry service for Bob-Lo was initiated from Gibraltar in 1984, one of the smaller boats was named the *L.R. Beattie*.

In November 1991, both the *Columbia* and *Ste. Claire* were auctioned off to top bidder Larry Spatz, a nightclub owner. Spatz envisioned turning the boats into boutiques or restaurants at the Detroit riverfront, next to his Baja Beach Club. Spatz was backed by a $235,000 loan from the Detroit Economic Growth Corp.

Spatz defaulted on the loan and never took possession of the boats that were stored at Nicholson Terminal and Dock, 10 miles down the river. Meanwhile, Detroiter William Worden established two foundations: The Steamer Columbia Foundation and the Steamer Ste. Claire Foundation. While sitting in storage the two vessels were designated National Historic Landmarks.

In January 1995, through the foundations Worden proposed to purchase the delinquent mortgage from the Detroit Economic Growth Corp. for one dollar and then foreclose on the outstanding loan so the groups could take possession of the boats at a federal auction. He promised to restore the *Columbia* and create a floating maritime museum. The *Ste. Claire* would be sold off and the profit would be split with the DEGC. The Growth Corp. readily agreed, and a district court judge ordered the auction. In addition, the judge ruled that Worden could use the delinquent loan amount to bid on the boats.

Only two people were present at the January 17, 1996, auction. And using the original loan amount—for which he paid $1—Worden outbid the private investor.

A buyer for the *Ste. Claire* came along in 2001. The private investor paid less than $100,000 and had the boat moved to Lorraine, Ohio, where it is currently under restoration. It had suffered from years of neglect and the Coast Guard worried that it would not make the trip to Ohio. It was discovered that rather then fixing a hole in its stern, someone had simply pumped it full with four tons of cement. With little fanfare, the Detroit—and now national—landmark made it to its new berth. The new owners plan to eventually have it once again open to the public with tours and possibly a restaurant. The Steamer Ste. Claire Foundation is supporting the restoration with fundraising efforts that include selling Bob-Lo merchandise and turning the boat into a temporary haunted attraction during October.

The *Columbia* has yet to find its fairy godmother. It is still owned by the Steamer Columbia Foundation, but the original plans to create a maritime museum have not materialized, and with every passing year the cost of restoration increases. The boat is now more than 100 years old and many fear it is rotting away beyond repair. The foundation originally planned on taking the proceeds from the sale of the *Ste. Claire* to restore the sister ship. However, when the *Ste. Claire* was finally sold, the money had to go to pay off debt for storing the boats. Since then, the foundation has taken a loan from the National Trust for Historic Preservation to continue to cover the cost of keeping the *Columbia*. That loan is now due and the foundation, still headed by Worden, has not raised the funds. The National Trust has threatened to hand over the *Columbia* to a New York group interested in preserving it.

THE SUMMER COTTAGE

While the park and boats were going through a chain of successive owners, life behind the scenes at Tresness continued to run its course. In 1984, after 42 years of marriage, Dorothy lost the true love of her life and partner in building Tresness. After the funeral ceremony, she brought Tress' ashes back to the island and scattered them at the back of the property, near the shed where he had spent most of his day working in or around. The following winter she retired, packed up her belongings and moved to Florida. But she never gave up the cottage and soon adjusted to returning each summer alone. Never alone for long, she had a constant stream of visitors, including a nephew and his family who would come every weekend. The days were spent on outside tasks or the occasional walk over to the park for ice cream (it was worth walking the extra distance for the real ice cream for $2 in a sugar cone by the Log Ride).

And then of course there was the cooking. No one visiting Dorothy ever left hungry. Offerings included everything from blueberry pancakes for breakfast to Hungarian feasts for dinner. On special occasions, she would take everyone across the river to Duffy's Tavern, an Amherstburg landmark, for her favorite meal—a plate of frog legs. Evenings were spent playing Aces until eyes couldn't stay focused. On days with no visitors, Dorothy would motor across stateside, visit with her mother, shop, do laundry, often staying until 10 p.m. Sometimes when she came back, she would be so exhausted it took everything she could muster to carry groceries up the hill. A wagon attachment for the John Deere mower given to her on her 72nd birthday became a most cherished gift.

In addition to family, Dorothy had a special relationship with many of the workers on the island. Many of them had befriended Tress, who had lent his expertise to many projects at the park; in return, they were more than eager to help with maintaining Tresness. In 1992 when Dorothy needed to replace the roof on the cottage, one maintenance supervisor sent a crew from his son's roofing company to do the job. The park plumber, Gil St. Pierre, was particularly close to Dorothy and would spend time tracking down parts for her boat and performing routine maintenance on the much-used John Deere mower. "I have more people than Carter has pills," she would quip about her island friends.

The park and longtime employees were a community, and Dorothy and her family became honorary members. Her diary entries include detailed conversations she would have with park employees, as well as other incidents:

"Russ (grandson) jumped out of bed around 2:30 a.m. and rushed to the park—someone was screaming for help (we thought). It turned out to be a party in the maintenance area. The park's new ground crew (hired outsiders) stay so late they can't get home so they stay overnite and were really livin' it up. They had booze, burgers and invited Russ to stay. He came home (Mon. a.m.) all loopy-legged around 6:10 a.m."

Despite the changes of ownership, it seemed as though the end of the park, and Tresness, was far from near. As she entered her 70s, Dorothy, too, seemed to be the antithesis of aging. Rather than shrinking, her frame physically got larger, and her strong traits of independence and stubbornness grew to mythic proportions.

One winter while in Florida, she made headlines as a pipe wrench-wielding grandmother who beat off two would-be attackers. The two men had forced their way into her home early one morning and had a confrontation with her grandson. Coming down the hallway, she saw the men punching him and holding him against the wall. She ran back into the bedroom, opened his toolbox and came out with the wrench. She struck both attackers squarely in the back before they fled out the front door. Out of the corner of his eye, Russell claimed he first saw his grandmother pick up a gun and fumble with it before getting the wrench. "If she had come out into the hallway with that gun, it would not have been pretty."

Those two men would not be the last to wish they had never crossed Dorothy Tresness.

YOU SEE THE PEOPLE WANT TO GO THERE

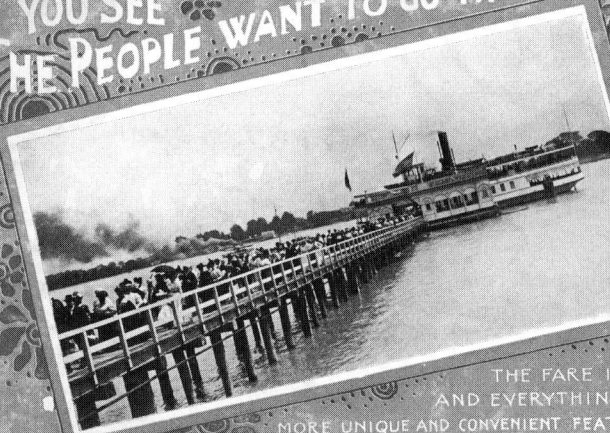

IT IS AN EASY MATTER TO SELL TICKETS FOR AN EXCURSION TO BOIS-BLANC PARK

THE FARE IS REASONABLE AND EVERYTHING IS UP-TO-DATE
MORE UNIQUE AND CONVENIENT FEATURES THAN EVER FOR 1900
DETROIT BELLE ISLE & WINDSOR FERRY CO.

FRED J. MASON EXCURSION AGENT

Aeroplane View of Bob-Lo Island Park, Canada.

Bob-Lo at the turn of the century. After signing a lease in late 1897, the Detroit Belle Isle and Windsor Ferry Company developed Bois Blanc into an excursion destination.

Copyright 1904 by the Rotograph Co.
A 3859 Bois Blank, Park.

Fun and Health FOR THE LITTLE FOLKS

We provide pleasure and comfort for young and old at

Bois Blanc Park

Season of 1900 opens June 10

Detroit, Belle Isle & Windsor Ferry Co.
FRED. J. MASON, Excursion Agent.

Simple pleasures. Early island excursions offered the promise of picnics, swimming and other family attractions. Some children, such as Warren and Louise Geoux (left), spent summers on the island at private cottages.

Building Boom. A block-house (left) built by the British around 1839 to protect Fort Malden in Amherstburg, has withstood the island's transformation into an amusement park. Amenities such as a cafeteria, roller rink and a "Women's Cottage" added to the island's appeal.

The Women's Cottage at

Bois Blanc Park

A building constructed in 1899 for the exclusive use of women and small children. No other Excursion Resort is equipped with a building of equal convenience and furnishing.

FRED. J. MASON,
Excursion Agent.

DETROIT, BELLE ISLE & WINDSOR FERRY CO.

Steamer Sappho

EXCURSION TIME CARD.

TUESDAY, JUNE 26,

For Lake St. Clair,

FARE, - 15 Cents.

SATURDAY, JUNE 30,

MUSICAL MATINEE

AND

Boat Ride to Lake St. Clair,

Music by the Detroit Opera House Orchestra.

FARE, 20c. Children, 10c.

Boat leaves foot of Woodward Avenue at 2 o'clock
sharp, for the above trips.

SUNDAYS:

For River Park Hotel, Wyandotte,

10.20 A. M. and 2.40 P. M.,

Fare, for Round Trip, 25c.

Busy waterway. By the early 1990s, five excursion boats—the *Promise, Pleasure, Sappho, Britannia and Garland*—were in service. A 1837 lighthouse (below) is still a landmark for boaters.

LIGHT HOUSE--BOIS BLANC--DETROIT RIVER

June 19-07 *Louise*

LAUNCH
"STR. COLUMBIA"
AT WYANDOTTE
SATURDAY, MAY 10TH, 1902
"STR. PLEASURE"
FOOT WOODWARD AVE.   1.30 P. M. SHARP CITY TIME
ADMIT ONE

THE DETROIT, BELLE ISLE AND WINDSOR
FERRY COMPANY WILL BE PLEASED TO HAVE
YOURSELF AND LADY WITNESS THE LAUNCH-
ING OF THE NEW BOIS BLANC PARK STEAMER
"COLUMBIA," AT WYANDOTTE, SATURDAY AF-
TERNOON, MAY THE TENTH, NINETEEN HUN-
DRED AND TWO, AT THREE O'CLOCK.

THE STEAMER "PLEASURE" WILL LEAVE
FOOT OF WOODWARD AVENUE AT 1:30 P. M.,
RETURNING AT 4:30.

Maiden voyage. As the park's popularity grew, small excursion boats couldn't meet demand. The 2,500-passenger *Columbia*—designed by architect Frank E. Kirby and built by The Detroit Shipping Company—was launched in 1902.

Bon voyage. With live bands, dancing and dining, the hour-plus boat ride to the island became an attraction in its own right. The 185-foot-long *Columbia* steadfastly ran the Bob-Lo route for 89 years.

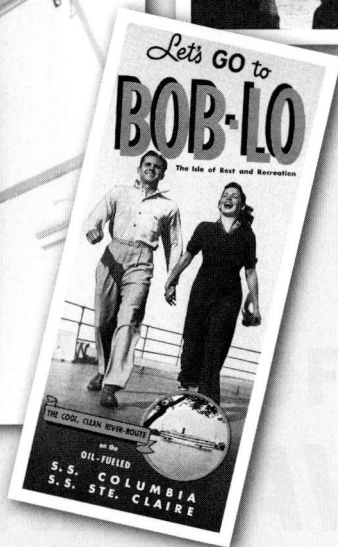

Sister ships. The *Ste. Claire* was put into service in 1910. The Ste. Claire and her sister ship the *Columbia* became known simply as the "Bob-Lo Boats." Captain John Sucharski (left) stood at the helm of the *Ste. Claire* from 1982 to 1991.

If the hat fits ... No trip to Bob-Lo Island was complete without a souvenir—and caps were a popular item for kids big and small.

Private property. In the early 1920s, Owl Cottage was the social epicenter of the island for a private resort community its upper-class residents called "Tanglewood." Visitors included the ladies of the Detroit Historical Club (bottom).

Louise Goux/Lovekin - Landlord

Solitude. Dorothy Tresness (below) and her husband, Orin, began building their own house next to Owl Cottage in 1955. Four decades later, Dorothy would have to fight to keep her island sanctuary.

Added amusements.
By 1923, the price of the boat ride had gone up to 60 cents. Owners began to use boat profits to improve the park. A new dance hall, the largest in North America, was made of steel and stone. Dancers were charged 5 cents a couple.

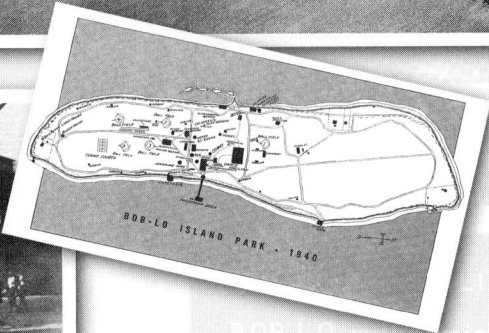

BOB-LO ISLAND PARK · 1840

Interior of Dance Pavilion, Bob-Lo Island Park, Canada.—2.

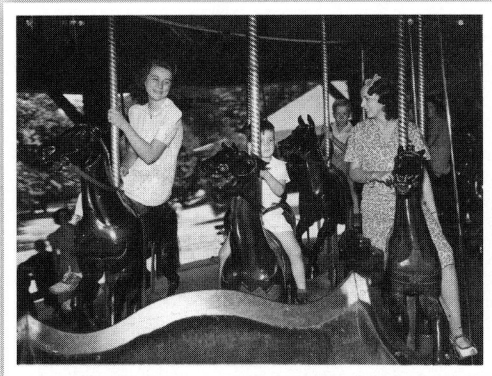

Rides and recreation. The first true ride on the island was a carousel that featured hand-carved horses from Italy. Horse rides and a golf course offering a gentleman's reprieve were soon added.

Golf Course, Bob-Lo Island Park, Canada.

On the tickets:

BOB-LO COMPANY
Good for one ride down-river during 1949 season, on any boat operated by the Company, between DETROIT and BOB-LO ISLAND on Sundays or Holidays, subject to conditions on reverse side.
ADULT (over 14) 65c
Federal transportation tax, if any, included in price. Not good if detached.
43114

BOB-LO COMPANY
Good for one ride up-river during 1949 season, on any boat operated by the Company, between BOB-LO ISLAND and DETROIT on Sundays or Holidays, subject to conditions on reverse side.
ADULT (over 14) 65c
Federal transportation tax, if any, included in price.
43114

Anticipation. Riders enjoyed scenic views and a visit with Captain Bob-Lo (above right), who worked the boats until 1974.

Bob-Lo Island CANADA

Arrival. When the Browning family began operating the park in 1949, they began making improvements. The family sold the park in 1979. In its heyday, nearly one million visitors were greeted at the park entrance each season.

Be Seeing you at Cool BOB-LO!

Dock ft. of Woodward. Sailings at 10 and 11 A.M., 2, 4 and 6 P.M. Moonlight at 9 every night but Mon. Extra Moonlight Sat. at 10. Round Trip $1.00; $1.30 Sundays. Children half fare. Sat. Island fee 10c. Moonlights $1.50.

Thrilling rides. The Brownings took a park that could have closed because of wartime economics and transformed it into a premier destination. Many rides—such as the Caterpillar, Tumble Bug, Whip, Scrambler and Comet—would operate for decades.

A parting gift. The brown mug with wooden handle is a classic example of a Bob-Lo Island souvenir.

The big leagues. Bob-Lo's status as a serious amusement park increased with rides like the Swiss Toboggan, Meteor, Rotor and a log flume ride. The mighty Sky Streek coaster is still running today in Guadalajara, Mexico, as "El Cascabel" or Rattlesnake.

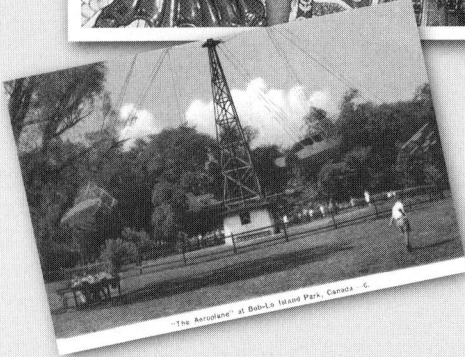

"The Aeroplane" at Bob-Lo Island Park, Canada — C.

SKY STREEK

Memories. Trinkets tell the story of Bob-Lo. For collectors and eBay traders, logos are key to identifying the time frame and shifting ownership.

LAST WEEK OF **BOB-LO**

SEASON CLOSES SUNDAY EVENING, SEPT. 12TH.

STEAMERS COLUMBIA AND STE. CLAIRE

Leave Week Days 9 a. m., 1130 and 3 p. m. Sundays 9130 a. m., 2 and 3 p. m., Eastern Standard, From Bates Street.

MOONLIGHTS EVERY EVENING EXCEPT SUNDAY AND MONDAY.

Columbia—Sunday Night Lake Ride and Concert, 35c

The Right to Refuse Any Person Admission to Boats and Park is Reserved.

The end of an era. In 1994, the amusement park was dismantled, the boats were left in dry dock and the island itself—minus a few parcels of private land—was auctioned off to a real estate developer.

Pieces of the past. Nostalgic for the park, Bob-Lo collectors shop on the Internet, at flea markets and antique stores.

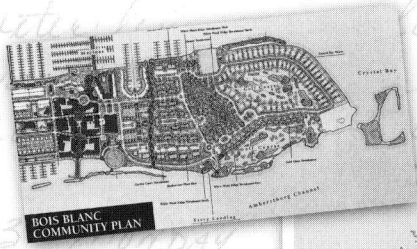

BOIS BLANC COMMUNITY PLAN

JULY 4, 1993 – SUNDAY –

July 5, 199... – ...DAY

July 7, 1993 – WED...

Standing her ground. Dorothy Tresness' stay on Bob-Lo Island has not been without bumps. She has endured battles over water rights, phone and property lines, the ferry, taxes— and even the right to walk down the road behind her own house. Her story is as turbulent as the history of the island itself. It is also far from being over. With the help of family, friends and even complete strangers, the octogenarian has steadfastly held claim to her private slice of island paradise.

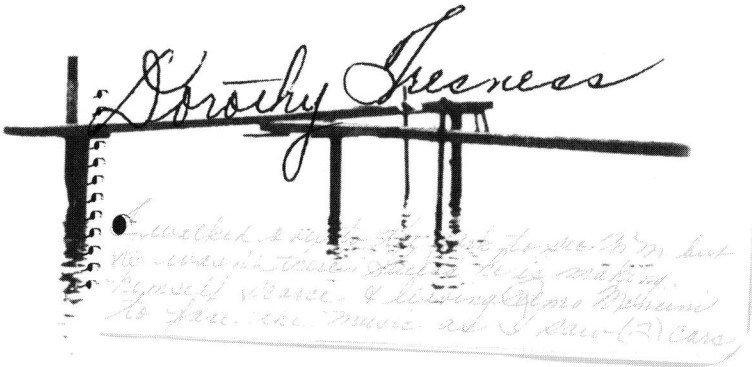

COME HELL OR NO WATER

"Each year I came up to stay, no matter what day it was,
I would just have to walk to the park and let them know I'm here
and they would turn on the water. Then it was business
as usual until September." —Dorothy Tresness

The Tresness property measures 100 feet wide and more than 800 feet deep, with expanses of green grass gently sloping down to the edge of the Detroit River. Both lots had running cold water since the 1920s, a benefit of the political prowess of the original owner, Leo Monahan. Leo was deputy city controller for the city of Detroit and had worked out a deal with the Detroit Windsor Ferry Company, the company that leased the island and began bringing people over to picnic. In exchange for dockage at the heart of downtown Detroit, the ferry company would provide water from the island's park area to the lots. It was a gentleman's agreement, Dorothy says, with nothing on paper. Back then things were done with a handshake and that was good enough, she says. Good enough to last nearly 70 years.

Having water from the park was never a clandestine act, and during the beginning of the 1990s Dorothy even paid to have the old lines replaced with more than 1,000 feet of new line, at IBC's request. Less than two years later, on July 4th weekend, 1993, new island owner Michael Moodenbaugh shut her water off.

Michael Moodenbaugh had purchased the island in an auction in March 1993. A tall, 6-foot 5-inch, impressive-looking figure, he was just 30 and something of a business dynamo. He came to Detroit from Seattle, where he was already successful partner in the revitalization of another amusement park, Enchanted Park, just outside of Puyallup. He saw a flyer announcing the auction of Bob-Lo and visited the island in February. One month later, he and partners plunked down $3.7 million and became the new owners. He had not even met the oldest resident of his new investment, but he was about to impact the rest of her life. With little hint of the battle that lay ahead, the 73-year-old widow began keeping a journal.

July 4, 1993 — Sunday

While eating breakfast, pump house engineer Gilbert St. Pierre came over and told us our water line was leaking and that he was turning off the water to let the puddle dry out so we could see where the leak is.

July 5, 1993 — Monday

Right after breakfast Robert, Art and Ronnie repaired the broken line. Broke same place as usual, right where it comes under the road where the heavy traffic goes over it. The reason why in 1991 IBC (park owners) asked us to install a new, more direct line. New line installed from our house to the park in 1991/92. With the bankruptcy problem their end was not completed, for one reason or another.

July 7, 1993 — Wednesday

G. St. P (pump house engineer) came over around 10 a.m. to tell me M. Moodenbaugh gave orders to turn off our water. I filled all the containers I had while Gil was walking back. At least they gave me enough warning to fill the bath tub, to flush the toilet and fill the water jugs and a 32-gallon trash can. Had 20 –30 minutes to fill everything I could.

I walked over to the park to see Moodenbaugh but he wasn't there. Guess he is making himself scarce and leaving Remo Mancini (president) to face the music as I saw two cars from the Ontario Department of Labor there—

probably about the accident Monday when one of the maintenance men was pitched off a cart. Figuring it would not be a good day to see Mancini I returned home.

July 8, 1993

It takes three times as long to get your work done (at least it seems so) when you don't have running water.

Gil came over today to see how I am making out. He told me Mr. Moody (Moodenbaugh) was after me. I don't know why, unless it's because he wants this property so badly. It seems he thought he bought the whole island.

Gil asked if I had a lawyer. I still don't want to start anything before I see either Moody or Mancini. They can't stand bad publicity and I don't want it pasted all over the paper about me being here. The fewer people who know about it the better.

GROWING FRUSTRATION

Thinking that if Moodenbaugh were aware of the grandfathered agreement he would honor it, Dorothy campaigned to present her case. Many times she made the half-hour hike to the former dance hall that had been converted into administration offices.

July 12, 1993 — Monday

Went to the park to see either Mancini or Moodenbaugh. Neither were available. Talked to Theresa (the secretary) and gave her the story of how we came to get water.

Moodenbaugh (they said) wasn't there—he was in Detroit/Wyandotte. Mancini was too busy. Personally think Mancini was deliberately evading me. I'm told he's "short of guts."

July 14, 1993 — Wednesday

No water—went to park again. Was told Mancini usually visits the marina first about 7 a.m.

July 15, 1993 — Thursday

Tried to see Mancini again today. Stopped at Maintenance Building and they called over to the main office. He wasn't available. I sure wanted to get the water on while Ron/Alice & boys were here from California. They're coming tomorrow and on Saturday the entire clan is coming plus my kids for a "sort of" small family reunion.

July 17, 1993 — Saturday

Had 24 visitors today plus myself—& no water. Thank God it was for one day only.

July 19, 1993 — Monday

Had a bad night last nite. Too much grease I guess. Shouldn't have eaten the skin on the chicken. About 2:30 a.m. I threw up and it was grease. Probably the potato chips didn't help either. After I heaved I felt much better & went to sleep.

Maybe the water situation is upsetting me.

July 21, 1993 — Wednesday

Went to park and tried to see Mancini again. NO LUCK.

July 22, 1993 — Thursday

Got up at 6 a.m. and went to the marina to await Mancini (by 7 a.m.) He didn't arrive by 7:30 a.m. Maybe late today as he partied with Moodenbaugh at the ball game yesterday.

Walked to main office—no one there. Started to go to little dock and wait for him there. Got to dock just as Mancini debarked and he just passed me by like a freight train passes a hobo.

Walked back to main office and his car wasn't there. So went to the pavilion (south side) where Mancini has little private office. I opened the door and there he was, finally. He wasn't going to talk to me. He said he was only following orders from Moodenbaugh but I stood in front of him, blocked his way, and gave him the story. I told him that as president of Bob-Lo Co. he was in administration and therefore answerable to the COO and was in a position to advise him of what was going on. When he saw that I wasn't going to shout and get hostile he did listen, but whether he'll repeat the story en tota I don't know.

The next time I go stateside I'll look up Moody at the Wyandotte office. I told him I didn't want to go to court. I don't want people to know I'm up here. I don't need the publicity and neither do they. They've had enough already.

I'm giving him every opportunity to not get unfavorable publicity.

When that confrontation still yielded no response from Moodenbaugh, Dorothy took another tactic and fired off a letter to Moodenbaugh. The letter was dated July 26, 1993:

Dear Mr. Moodenbaugh:

Because I have been unable to contact you personally and knowing how busy you are, I am writing in an attempt to clarify why Bob-Lo furnishes water to our cottage "in the bush."

Years ago when Leo Monahan was deputy city controller of Detroit he was instrumental in securing dockage rights for Detroit Windsor Ferry Co. at the foot of Woodward. In exchange for this, DWF Co. offered to furnish water to our seasonal dwellings.

In the early '30s there was a dispute and the matter was taken to court where Bob-Lo was informed they must continue to furnish our water. They could, however, charge for the water; but that would necessitate maintaining the line. Because we consume a minimal amount of water it was their decision to continue furnishing water at no charge.

The above can be verified by your solicitors. It is a matter of court record and I do believe you will find Ontario law does not permit withholding a life-sustaining commodity such as water.

I assume Remo Mancini informed you of our July 22 conversation regarding the above.

If more information is needed to expedite your turning on the water, please contact me as your many business commitments make it difficult for me to reach you.

Yours very truly,
Dorothy Tresness

July 29, 1993 — Thursday
Faxed letter to Moodenbaugh today. It was received at 11:59. Teri (niece) spoke to a Dora Ann (secretary) at 12:55 and she said Moodenbaugh was reading it then.

GOING PUBLIC

Another five days went by and still no response from Moodenbaugh. As Dorothy was approaching one month without water, word started getting out through the Canadian press. While some encouraged her to fight her battle through the media, in her journals she contemplated the ramifications of "going public."

August 4, 1993 — Wednesday

Heard on radio Moody threatened to close park if Canadian Coast Guard did not give approval for him to use his U.S. boats to transport passengers from A'Burg to the Island. He's also asking the Gov't for the right to use U.S. help for special events. I can't believe the cheek of the man. Guess he has more nerve than a barrel of monkeys. Here he's asking Canadian concessions and won't even turn on my water. I guess I just don't have the push.

Should stop trying to reach Moody and Mancini and go to the media; but I hate publicity. Lord knows how many I'll have gawking if and when this hits the public. Well, it might serve some good—the boats may slow down.

August 5, 1993 — Thursday

Tried to call "The Reeve" (Amherstburg mayor) Gibbs again. Got the machine—again.

Was going to call the *Windsor Star* when some park employees pointed out Tom Aubin from CBC-Radio. He conducted a short interview and said he was going to call Reeve Gibbs when he returned to the office.

About 3 p.m. David Morelli, reporter from *Windsor Star* pulled up at the dock and conducted an interview and took pictures.

August 6, 1993 — Friday

Walked to the park at 11 a.m. to get a couple gallons of water. Wonder how long it'll be before Moodenbaugh finds out I'm walking over to get water.

Noon — 1:30 p.m. The dogs were raising Cain because a big 27-foot cruiser pulled up at the dock. It was a TV crew from CBC. Lady reporter Josee Guering. They had been "escorted" off the island so had to hire transportation here. They said Morelli had his story on page 1 of the *Star*.

Right about supper boats started tooting and giving the thumbs-up signal when they went by & 4 different boats left water. The first three just dropped a gallon and left. Then another boat landed and a man got out after putting one case of water on the dock and carried up another. He gave me his name and phone number in case I should need help.

I never knew there were so many caring people in this world.

August 7, 1993 — Saturday
Heard on radio today that Canadian Coast Guard has given approval to Moody to use the U.S. boats in Canadian waters, I'd like to see a headline "GOVERNMENT CAPITULATES TO BOB-LO BUT BOB-LO RE-FUSES WATER TO 72-YEAR-OLD WOMAN."

August 10, 1993 — Tuesday
Went to Bob-Lo office in Wyandotte to try and see Moody. His secretary said she'd set up an appointment and let me know. I told her I didn't have a phone most of the time. She said he usually spent time on the weekends so I told her to have the office send one of the workers down to tell me when to meet and I'd be there.

Many longtime employees of the park continued to be Dorothy's allies and served as her pipeline to the communications (official and unofficial) coming from park management. They stashed water bottles for her and de-livered them in park vehicles. One security guard at the marina let her slip in and use the showers. She learned from sources in the maintenance depart-ment the rumor that Steve Langdon of the Canadian Federal Parliament had been in touch with management and told them to turn on her water. They also gave her a copy of the Amherstburg paper in which Moodenbaugh was quoted, "You don't get to be a millionaire by giving away water free."

Things remained relatively quiet for a couple of days, but by the end of the week, Dorothy went from low profile to front and center in the media circus.

August 11, 1993 — Wednesday
Found a note under my door from Rob Musial, *Detroit Free Press*. Called him and left message that I was here.

August 12, 1993 — Thursday
Rob Musial, DFP, came over and took several rolls of film while interviewing.

August 13, 1993 – Friday
Musial's article hit front page of the paper. Around 11 a.m. (2) boats—one with TV2 and the other with TV7.

Huge cruiser tonite with (8) gal. of water.
Must have been on news. Not having hydro (Canadian term for electricity) I don't see TV.

August 14, 1993 – Saturday
Operation "Water Jugs" today. Some Grosse Ile Yacht Club members came over on *The Duck Factory* and (2) other boats.

Had TV7, TV4, TV50 and lots of people—Lucia and Stan Sawka, teenage children—all carried water. They left around 1 p.m.

Got batteries for the 7-inch TV and saw "the fiasco" on the 11 p.m. news—TV7 and TV2.

It made for good press—a sweet old grandmother being bullied by a big corporation.

Detroit and Canadian readers and viewers were soon treated to images of Dorothy in her pink tennis shoes, scooping water out of the Detroit River. She would wake up in the morning to find her dock lined with gallon jugs of water left by boaters and would pull the bottles back up the hill 500 feet to the house using a little blue wagon. Her brother, who lives in the Manitoulin Islands, saw his sister's plight on TV. The story had gone national and beyond, appearing on CNN. The water donations kept coming all summer. One day someone even brought her a large purified water dispenser, the kind found in offices. But while her 15 minutes of fame consisted of one summer, the water problem would remain unresolved for years.

In an article in *The Detroit News*, Moodenbaugh claimed that his insurance company made him do it, citing that he would not be covered if anything happened to property owners because of the island's water. In the same article Dorothy was quoted saying she would "hang in here if I have to haul water until I'm 90." The Canadian press painted him as an American bully. Rumors of company boycotts with both American and Canadian organizations refusing to hold their annual picnics on the island ran rampant. Dorothy hired a lawyer, and met clandestinely with him on park benches in Amherstburg, knitting

Christmas stockings while discussing the situation. She met the Amherstburg mayor in a restaurant. Over bread and black coffee, he assured her he would talk to Moodenbaugh.

Confident she was doing everything she could, Dorothy readied to return to Florida to wait over the winter for good news from her lawyer. Her last diary entry for the summer of 1993 read:

"Went to the park to use the phone to call lawyer. He was with a client so I killed 45 minutes by calling Gil St. Pierre. He has been laid off and will be terminated because he is too old and too slow. Moody is nuts. Gil has been the pump house engineer since as long as I can remember and no one knows the lines around the island like he does. Now he won't be able to finish my line … such a sad ending for him, hopefully all is not lost for me."

BIG PLANS

Moodenbaugh, who seemed to consider the water issue over and done with, was directing all of his energies toward the first season at his new park. Brimming with plans to turn Bob-Lo into an even larger-scale amusement park, he spoke of building a new marina, new rides, a wave pool and hosting concerts featuring top pop and rock acts.

Considered an outsider, one of Moodenbaugh's first tasks was to make himself known and gain support from the powers that be in the local Amherstburg government. The Canada connection proved to be turbulent; his unfamiliarity with Canadian and maritime laws did not enable him to move as fast as he would have liked. He had trouble with boats and the issue of where and how U.S. and Canadian vessels could be docked, and threatened to close the park. He had six boats, but only two were registered in Canada and authorized to make runs from and to the Amherstburg dock. This caused back-ups, with people waiting as long as two hours to take the ferry to or from the island. When he threatened to close the park and handed layoff notices to all 860 employees, the majority of whom were Canadian citizens, Canadian immigration agreed to let U.S. crews run the four U.S. boats between the mainland and the island.

Things seemed to be shaky, but the new blood apparently did re-energize the aging park. In his first 45 days, Moodenbaugh had hired a staff, grounds crew and 800 summer employees to get the park ready for its 1993 Memorial Day weekend opening. Changes he made right away included a Trinidad Tripoli steel band at the dock welcoming visitors, marching bands and a new mascot, Bob-Lo Bear. Moodenbaugh entered into new contracts with name-brand foods such as Pepsi, Colombo frozen yogurt and Kraft to replace local food vendors. He also secured a liquor license for the island and opened bars and restaurants.

The marina was renovated with a bar, sandy beach and even a pizza delivery service right to the boats. Attendance during the Fourth of July weekend set all-time single-day and weekend records. The 138-slip marina filled to capacity and more than 75 boaters were turned away. The popularity continued throughout the summer and in fact, attendance records showed that the season Moodenbaugh opened, park attendance doubled from 200,000 in 1992 to 400,000 in 1993.

Yet, midway through the summer there was unrest among the partners. Friend Jeff Stock left the partnership they had formed (Omni Properties Ltd.) leaving Moodenbaugh holding the bag with silent investors the Benaroyas, a benevolent Seattle family. It had been the Benaroyas' company, Northern Capitol Corporation, that had fronted Stock and Moodenbaugh the $3.7 million. In exchange, the partners were to repay the loan and then split the profits 50/50 with the Benaroyas. Moodenbaugh was given a management contract to run the park. It was an arrangement similar to the one reached with a park in Seattle, which had turned an impressive profit in the first year. The Benaroyas were apparently not pleased with the rate of return at Bob-Lo. Although it had only been a matter of months, they indicated to Moodenbaugh that they were ready to call it quits on the Canadian amusement park and that his management contract would not be renewed come September 29. Moodenbaugh began to search for potential new financers.

END OF AN ERA

On September 23, Moodenbaugh and an employee got into Moodenbaugh's 1987 Ford Bronco II and took a short trip to Toledo, Ohio, to promote an upcoming concert on the island. It was after midnight when they decided to return home. Merging on the ramp to I-280 in Toledo, with Moodenbaugh driving, the vehicle went off the right shoulder and flipped four times, coming to a rest on its roof. Moodenbaugh's body lay on the pavement, 30 feet from the truck. He had a broken back and was in a coma for weeks.

Moodenbaugh survived, but his hold on Bob-Lo did not. September 26 was the last day of operation for the park. On September 30, 1993, Larry Benaroya was granted control by the courts and he officially closed the island. As Moodenbaugh lay in a coma, plans were already in the works to dismantle the rides and sell off the park piece by piece, thus ending a long and cherished chapter of Detroit history. Events were canceled, including an Octoberfest, a Haunted Island attraction and a Halloween concert that was to have featured the band Electric Light Orchestra. Some 12,000 schools were notified that the Spring 1994 free day on the island they had signed up for would not take place. While some loose ends were being tied up, no one gave a thought to the future plight of the widow "in the bush"—still living without water.

With a long recovery ahead of him, Moodenbaugh went back to Seattle and was silent—until three years later when he filed a lawsuit in July 1996 against his former partners, charging them with breach of contract, conversion of assets and defamation of character. Shortly after the filing, a letter that was posted on the Internet and printed in *The Detroit News* was Moodenbaugh's first and only attempt to explain to the people of Detroit what had happened with the park:

"Read your story on the auction at Bob-Lo. Pretty sad that my partners at Bob-Lo let a personal dispute wipe out a 96-year legacy. This is Michael Moodenbaugh the former owner at Bob-Lo. Had I not been in a coma for 75 days I would have stopped the Benaroyas from ripping the park apart. The trial will bring out the truth about what really happened. After the trial I intend to publish the real story."

The battle never went to court and the story seemingly ends there. There was an out-of-court settlement with Moodenbaugh collecting an undisclosed amount from Stock and the Benaroyas. But there was no chance that he could ever have a hand in the island again. Bob-Lo the amusement park was gone. The island was put into receivership and auctioned off in 1994. For the first time in 97 years, it was sold not as an amusement park but as a piece of real estate with virtually unlimited potential. The auction brochure touted its potential for development of casinos, condominiums, a resort and conference center. Calling it a "City-In-A-City," it described the infrastructures—including a power cable that ran underwater from the Canadian mainland, an on-island back-up generator system and a self-contained water supply and sewage treatment system. The winner could have it and do what they will—minus the property owned by the old widow and the only full-time island residents, a family who owned 17 acres on the northern tip of Bob-Lo.

The man who stepped up to the plate was John Oram.

MODERN (IN)CONVENIENCES

*"He invited me to the press conference when he announced he
had bought the island. He introduced me and was patting my back ...
little did we know he was patting us to find the soft
spot to put the knife." —Dorothy Tresness*

John Oram was a 41-year-old businessman from the Detroit area with diverse holdings in real estate, a car stereo business and two hotels near Detroit's Metro Airport when he stumbled upon his next venture. Hearing that the island was being sold following the Moodenbaugh accident, Oram inquired about buying the Nightmare roller coaster for his son. Not only was the roller coaster for sale, but also the entire contents of the park and the land on which it stood. With visions of creating a year-round resort-style community, Oram ended up purchasing the island for $2.5 million. The price tag was quite a change from the island's heyday, when it was sold for $20 million in 1988.

The new owner and his plans were unveiled to the Detroit and Canadian media at a press conference at Windsor's Dieppe Center on August 10, 1994. In addition to Oram's wife, four children and extended family, government leaders from Windsor, Amherstburg and Detroit were on hand as he outlined his plans for the development. One audience member in particular, wearing pink tennis shoes, was listening very carefully as the new owner made promises to maintain the natural beauty of the island. Prior to the press conference, Oram had taken Dorothy on a picnic, shared his plans with her and assured her that she would have water.

Oram's purchase included not only rides, but also two tugboats, one barge, tractors, an ambulance, 1,200 picnic tables, 800 park benches and 2,500 folding chairs. He also inherited the Amherstburg dock and parking lot that sat on a 92-acre parcel on Highway 18 at the edge of the township, a separate freight dock and outbuildings, and office space on the U.S. side of the river. The detailed auction list also included funnel cake mixers, an 800-gallon water wagon, office furniture, drill presses and 300 pairs of used roller skates.

Oram's grandiose plans included a new nine-hole golf course, restaurants, fudge shops, a petting zoo, a dinner theater and a casino to attract weekend boaters to the island as well as year-round residents. Within a year the first 52 residential waterfront lots ranging from $90,000 to $270,000 were up for sale. One of the lots was where the Owl Cottage stood. When Louise Lovekin signed over the papers, Oram wasted no time in bulldozing the house. As the machinery chewed it up from the outside in, lumber and shingles came crashing down along with a century's worth of memories. Watching from her house next door, Dorothy was crying so hard she couldn't hold a camera steady to take pictures. To this day when she looks out her window, she doesn't see the mammoth new house next door; she sees the familiar octagon-shaped roof and hears the porch door swinging open, the sound of kids racing each other down the hill toward the water.

A RIDICULOUS OFFER

By 2002, Oram sold more than 80 homes and began building condos. But there was one parcel he could never purchase.

For one summer, Oram allowed the cottage to hook up to the island water treatment plant, but after he began development, Dorothy was once again cut off. He then offered to buy her out for $125,000 Canadian—what she calls a "ridiculous amount." Her property's value is high, perhaps incalculable because of the water rights. Hers is the only lot that includes part of the Detroit River in its property line. This gives her the right to dock a boat on her property, something the mansion owners cannot do. They must pay to use the island marina, operated by Oram. She refused his offer:

"I am not interested in selling at this time.

"I despise the hubbub, the conflict, and the upheaval we are experiencing at this time; but this is 'progress' and there is not much that can be done about it. At 76 years of age all I really want is to continue to enjoy my summers on Bob-Lo Island and die in peace! Then my children can sell or 'fight the battle.' "

Shortly thereafter, Dorothy received a letter from Oram stating that the charge to connect to the water and sewer would be $91,680 Canadian.

Oram continued to take actions to ensure that the cottage would not get water, or any other utility for that matter. He proceeded to build up around her and placed a meter of easement around three sides of her property, effectively boxing her in. This easement was considered private property and prevented the electric company or water company from running lines to the cottage. The new sewer lines stop just a few feet short of her backyard, and when they were installed, contractors left a ditch at the edge of her property, over which she now had to drag her trashcans for pickup. Oram also planted rows of evergreens on the easement, ostensibly to hide her property from view. In doing so, he blocked her driveway from road access. Finding it difficult to part the trees and jump the ditches at her age, Dorothy requested that a culvert be built and took her argument to the mayor of Amherstburg. When her 2001 property assessment skyrocketed, Dorothy also made it a point to let the Ontario Property Assessment Corporation know of her situation.

Citing her property "virtually useless" she wrote the following appeal:

- "Island developer John Oram has a 3 meter reserve around our property. He will not grant an easement to Ontario Hydro to cross this reserve. Therefore, we have no hydro.
- We have no running water.
- We are not permitted to use the ferry even though my grandson profered the (exorbitant) fee, in cash, to Oram's agent. It was refused. Access

to our property must be by private boat. This means any building and repair materials must be transported in our own (small) boat.
- Oram has planted evergreen trees around our property. Even though our
present home has been here since 1960, he does not want prospective
buyers to see our Canadian farmhouse among his upscale Victorian homes.
- I am 80 years old and no featherweight; should I ever need help EMS
would be unable to bring in an ambulance to administer care. Oram's evergreens are planted in his 3 metre reserve, blocking access to our drive.
Further, the Police and Fire Departments cannot access our property except on foot. According to the Fire Chief this is hazardous."

Her assessment was promptly reduced 75 percent. But there was no response on the water issue.

MAKING DO

Continuing to make the best of the situation, Dorothy's industrious grandson built a Rube Goldberg-like contraption that gave her running water. A
100-gallon tank perched in the attic routed water through the floor, down
the wall of the kitchen and into a plastic tub in the sink. The water inside the
holding tank came from her new neighbor, who ran a hose from his $2 mil-
lion house across 170 feet of property to the cottage. She used the water to
wash dishes and hands, after treating it with bleach. But it is not enough to
take a bath or to flush the one toilet in the house. Urine was collected in a
"potty seat" perched on the toilet, like those used to train toddlers. If nature
called for more than that, there was always the outhouse out back.

It has that traditional outhouse look, a brown box with a hinged door on
it, and Dorothy doesn't think twice about venturing out there during the
night. It also has a rather large window so one can truly sit and commune
with nature; but, as is her approach to most things in life, Dorothy is a get-in,
get-out kind of gal. Tress used to spend a lot of time in there, as most men
do, she says. Of course, back then there was more to see than condominiums.
There were black squirrels, deer and rabbits, pheasants, quail and other types
of fowl. Today, the prevalent wildlife is Canada geese that fly over so low
their shadows cover an area three times as big as their bodies, their constant

honking echoing off the water, sounding like rush-hour traffic has suddenly converged on the island. In one journal entry, Dorothy counted 76 geese feasting on her lawn.

The outhouse burned down once and was rebuilt before it even stopped smoldering, she says. She laments that her grandson relocated it recently—putting it closer to the house and farther from the main road, where prospective mansion buyers would see it as they toured the island with Oram.

The house was wired for electricity when it was built, but was never hooked into the lines from the park. The island was originally lit up in 1906, powered by generators until Ontario Hydro installed the cable in the early 1960s. At that time, the Tresnesses could have paid to run lines from the park to the property, but even then it would have been an expensive proposition and one that they just didn't see the need for at the summer cottage. Once Oram created the easement, Ontario Hydro refused to cross it without a court order.

Living without electricity was not particularly disconcerting to Dorothy, although she would often wonder what would happen should her 1937 Servile gas refrigerator stop working. It worked better than a freezer. Lacking a thermostat, it would be continually "on," and freeze everything solid. Lacking in modern conveniences, there was no such thing as fast food at the cottage. She often had to wait for her milk to thaw, and a hankering for a hamburger would not be fulfilled for two days.

Bottled gas also powered the overhead lights, reminiscent of those found in Amish homes, cheesecloth covered and emitting bluish light. The lack of electricity led to a décor theme that may be a bit unusual but served a useful purpose. Battery-powered flashlights of every size and color were strategically placed so there was one within arm's reach from any point in every room, hanging on hooks, sitting on shelves, on ledges above every door and attached with magnets to the side of the appliances. The flashlights came in handy, particularly on rainy days and for reading after dusk. Dorothy read 30 to 40 books every summer. In her journal, she tracked when she started and finished the books. She also developed her own critiquing system with a series of checkmarks and symbols. In one summer, entries included "The Chancellor Manuscript" (Ludlum); "Clear and Present Danger" (Clancy);

"Jurassic Park" (Crichton); "A Necessary Woman" (Van Slyke) and "Time & Tide" (Fleming). Writing in the margins of the spiral bound journal a typical entry appears:

"'Magic Hour'—Susan Isaacs mystery 257 pages 7/1, finished (450 pages) 7/2-3 (12:30 a.m.)"

Of John Grisham's "The Firm" she wrote: "The best book I've read in years. Couldn't put it down! Read until 2 a.m. and got up at 6 a.m. and finished it."

On special occasions she would get out the 7-inch battery-powered television. On 9-11-2001, Dorothy heard about events from a telemarketer. "She said, 'Isn't it terrible what happened in New York?' And I said, 'What happened in New York?' Then I turned on my radio."

She kept a battery-powered radio and read newspapers from Amherstburg, Florida and Michigan. She loved to work cryptograms and cryptoquotes to "keep the brain active." When it finally got too cold and too dark, Dorothy would close up for the winter and go to what she now calls home in central Florida. She was a snowbird but she didn't fly. She drove up from Florida and left her car on the Detroit side at her daughter's house so she could do her own shopping, doctor's appointments and visiting.

The time she spends on the island has gotten considerably shorter; she often waits until the warm months of June or July to make the trip. One recent summer, her grandson offered to fly down and drive her back but by the time he had bought a ticket she was already en route in her purple Escort station wagon, with her two dogs and an assortment of books on tape. She always stayed until her birthday, September 27, and then made the drive back. While she was gone, her grandson became the unofficial caretaker.

Tresness was never a year-round home. When they lived across the river in Michigan, occasionally Tress would come over in the winter and discover evidence of visitors. A burn hole remains in one ceiling tile from when duck hunters taking shelter built the fire in the fireplace too high.

Dorothy, a U.S. citizen and Canadian "guest resident" dutifully paid her taxes and in return, her garbage was picked up once a week. The summer cottage still had no electricity, no water, but it did have a phone and a new address. Ironically, Dorothy lived on Oram Drive.

24 HOUR YEAR ROUND
FERRY SERVICE

EBB AND FLOW

*"I could do without the telephone. I went without it for
all those years, but I guess when you're 82 years old people
just want to know if you're still breathing.*

*"The telephone company, Ontario Bell, came in on August 18, 1997, and I
got me a phone! They came and put it in and Oram ripped it out. But you
should have seen how that line was when they put it in. They promised
me they would put my phone in and the day they put that phone in it was
raining something awful. Now in the United States if they were going to
put in a phone and it was raining cats and dogs they wouldn't come out;
but they came out there in the rain and they had to go 200 feet up the street
to attach me in to Bell's service box. Then they came down in front and they
hung it in the trees, just draped the wires across my trees, across my gutters
and down and in the house. Then within the week they came back out and
dug a trench and put the wire out and down the street but they attached it
to the house next door. Well, that was where Oram's vice-president lived
and she gave them a fit what for. She gave them a fit every day until
they came back and rearranged it." —Dorothy Tresness*

Dorothy not only recalls the exact date, she even remembers the names of the men who came to put her phone in. Through the pouring rain, she could just make out the yellow slickers of the men bobbing up and down in a boat laden with coils of insulated wire. They were there to deliver on a promise made to the feisty old lady who was right out there with them in the rain as they ran the line.

Draping wires across the property would have never have been up to code in Canada or the United States, but this was a special circumstance. Ontario Bell had decided it was an endangerment to the older woman's life to be on the island without a phone. When Oram confronted them for crossing the easement and ripped the line out at the service box, the men told him if he dared to cut off this woman's phone Ontario Bell would not provide service to any of his residents.

Prior to that, family and friends had two options of contacting Dorothy. Those who had boats could drive over and hope the water level was high enough to pull up to her property, or they could try to call on the mobile phone. The mobile phone of yesterday was a far cry from the technology of today. As Dorothy says about the bulky square apparatus, "It was never any good because it had to run off a 12-volt battery which was often dead." It looked like something seen in the war movies—a big black box. And in many ways, she was like a general, directing the moves of the island traffic, the matriarch of the island. A dead battery was a nuisance, because with no electricity, there was no way to re-charge it. She worked out a system with family that her "calling hours" would be between 4 and 6 p.m. each day. Friends and others would often relay messages through her daughter. As a U.S. resident, she was required to check in with Customs by phone to let them know when she was staying on the island. In earlier days, she would walk to the park marina or laundromat and use the phone. Once Oram bought the marina, however, he revoked that privilege.

MAROONED

For the new Bob-Lo Island residents, a ferry owned by Oram became their mode of transportation. The ferry was necessary because there is no bridge crossing the 875 feet to Amherstburg, and also because even though the residents owned their homes on the water, they did not have water rights. So they were forced to pay to dock their own boats at the marina, which was only open during the summer. The ferry remained the only way to get a car back and forth. For anyone working in Canada, or simply needing groceries, the ferry was a lifeline. The island had no amenities, only a small restaurant and convenience store that was open sporadically. The major shopping and entertainment district that Oram promised never came to fruition. Although it ran 24 hours a day, seven days a week, year-round, Dorothy rarely utilized the Amherstburg ferry as she always had her own boat and left her car on the U.S. side rather than Canada.

She bought boats like the average person buys jeans, and knew the workings from the inside out. In 1988 she took a Coast Guard course because completing it entitled her to a sizable discount on boat insurance. She then joined the Coast Guard Auxiliary and still crews for Flotilla 1508 in Hernando Beach, Florida, during the winter months. She volunteers some 182 hours each season.

"My brother said, 'I sure would hate to be out there and see some little old fat lady coming to rescue me.' Well let me tell you something, when they see this little old fat lady with a line they're damn glad to have me," Dorothy says.

The purpose of the Auxiliary is to help stranded boaters as well as educate people in boating safety. Auxiliary patrol boats typically spend six hours at a stretch looking for anyone that has broken down on the water or is in need of assistance. The crews perform hundreds of rescues but not every day is eventful. "Sometimes we just have a ball," Dorothy says. "Although I did have a skipper drop dead at the wheel once. We had to leave the boat we were towing. I'm not a coxswain so I'm not allowed to run the boat, but I could have if I had to."

She always kept her own boat in storage for the winter in Michigan. When she returned in the summer of 1999, she decided it was time to upgrade and went to an annual boat show on Lake St. Clair.

Knowing exactly what she wanted, she approached a salesman. He looked at the old woman with disdain and tried to give her the brush-off. His mistake, as she had a brown paper bag tucked in her purse with the cash to buy the boat. Striding over to the next marina, she ordered a new red Larsen, marched back to the first salesman and waved the purchase receipt—stamped "paid in full."

She gave up that boat in 2001 at the age of 81 and decided there will be no more. Successive years of low water levels in the Detroit River have left her dock sticking out more than 6 feet above the water, too high for Dorothy to climb up and down from. In order not to run aground, any boat approaching her property now has to be tied up 50 yards from shore. Plus, as she was spending less time in Michigan, she was spending more money to keep the boat in storage. "And I just couldn't train my two dogs to be a good crew," she jokes. In perhaps what was the only time in her life that her practical side overruled her independent nature, she relinquished her title as a boat owner.

When she sold her boat, Dorothy's grandson took an envelope of cash to Oram's office to pay for her to ride the ferry for one year. Oram refused to accept the money, saying she was not part of his community and would never be allowed to ride the ferry. She was left to wade out into the river, carrying her belongings in black trash bags to her grandson's boat.

While Dorothy suffered under Oram's monopoly, she would no longer be the only one. Soon all residents would face a similar situation as trouble began to ripple once again around Bob-Lo.

In early 2003, the island's other private residents were also denied use of the ferry. The couple, who worked in Canada and had two school-age sons, had lived year-round on Bob-Lo for 10 years. Their land had been the second private plot excluded in the original sale to Oram. When he learned that they filed an application to sever their 17 acres into multiple buildable lots, he gave them a 12-hour notice that the ferry would be off limits. The City of Amherstburg refused to step in, so the couple sued. Citing an old law on the books that deemed children could not be denied an education, the judge decreed that only the children be allowed to ride the ferry, and only during the school year.

Dorothy joined the couple in a second lawsuit to again try to force Oram to give them all full access to the ferry. Their lawyer's plea to the Amherstburg City Council to take over the ferry service fell on deaf ears. The mayor stated: "Mr. Oram owns the ferry. Mr. Oram runs the ferry. The municipality takes no responsibility for passenger transport to and from the island." But soon the city would no longer be able to ignore the effects of letting Oram run the island.

THE FINAL BLOW

On May 5, 2004, Bob-Lo Island was placed into receivership by the Canadian government as Oram had defaulted on more than $19 million (Canadian) in mortgages, construction liens and back taxes. Oram's brother, who had helped him purchase the island, accused him of diverting corporate earnings, witholding payments and forging company documents.

The first outward indicator that Oram was in serious trouble had come that March when he took the ferry out of service. Suddenly the residents were left to find their own way on and off the island. Neighbors began to share boat rides and posters were put up announcing scheduled runs to the grocery stores. The more ingenious folks bought kayaks and canoes. Collective letters to the editor were written and motions were made at city council meetings. Finally service was restored by the court-appointed receiver, only to be again suspended for most of the summer. This time, the receiver claimed mechanical problems.

The charm of island life was wearing thin for many of the residents, and by the end of the summer all but seven homeowners had fled to hotels or homes of friends and families on the mainland. Over at Tresness, it was business as usual.

"These folks are getting a taste of what I've lived with for years," Dorothy commented.

Eventually ferry service was restored. On its first reinstated trip to Amherstburg, no one questioned the elderly lady in line in her Escort station wagon. For the first time in years, Dorothy went grocery shopping by car.

That ride marked the second victory for the woman who for so long was denied island services.

The first victory had occurred when she had taken Oram to court in 2002 over the water issue. That Christmas she received a call at her Florida home from her lawyer. The Ontario Municipal Board had granted her the right to cross the easement to hook into the island's water and sewer. Ready with the $21,000 (Canadian) necessary to obtain the permit, Dorothy returned to the island in the summer of 2003 confident that she would have running water by season's end. When she arrived, however, she found that paying it would do no good—Oram would not allow the equipment needed to dig the trenches to be brought over on the ferry. A failed attempt to hire a private barge to bring the equipment over only made Dorothy more determined. "The tide has changed," she declared. "I can feel it in my bones."

EPILOGUE

By the time she returned in the summer of 2004, things had indeed changed. In this David vs. Goliath story, the menacing giant had been run off his own island. The receivership and legal turmoil had generated media coverage, and the local Canadian paper revisited the woman it had featured on its front page 11 years earlier. When it was found that her living conditions had remained relatively unchanged, Dorothy once again made headlines. The reports drew the eye of an Ontario Hydro employee who had been aware of Dorothy's situation for many years. Campaigning to provide electricity to the cottage, he was told repeatedly by Ontario Hydro corporate lawyers that they could not run lines across the easement. With Oram now removed, the man tried once again and finally got the green light to service the house in the bush.

On October 4, 2004—for the first time in 64 years—Tresness was aglow with electric lights.

As she celebrated another victory and her 84th birthday over a meal of frog legs at Duffy's, long-term plans involving the ownership of the island and the ferry remained unresolved. Media reports pointed to a Windsor-area builder and former business partner of Oram as the next man in line. As the builder of most of the homes and roads on the island, he was the largest creditor (owed nearly $11.5 million Canadian). He was asked to submit a reorganization plan. Trying to hang on to the island, Oram also submitted a plan. It called for selling part of the island for $1.5 million and borrowing another $5 million. Lawyers scoffed at the notion of letting Oram borrow any more money. Oram appealed to the courts, adding he would work for "$1 a week" but could not avert the receivership. The court accepted the builder's reorganization plan and placed responsibility for the island in his hands.

Yet again, the tide of ownership had changed on Bob-Lo.

Braving the cold chill of autumn that year, Dorothy stayed on the island through October, determined to see Oram have his day in court. When he walked into the courtroom, she was seated right up front—pink tennis shoes on. Her presence signified the final victory in what had been more than a decade of fighting. She had endured battles over the water, the phone, the ferry, property lines, taxes—and even the right to walk down the road behind her own house. For her, seeing Oram in court was a closure of sorts. He had the appearance of a broken man, she says—"about as sad as a one-armed paper hanger with an itch."

Before returning to Florida, Dorothy paid the $21,000 (Canadian) for the water permit and the new owner promised to dig the trench the following spring. The residents asked her to join the homeowners association, but she refused.

The story of Dorothy Tresness is as turbulent as the history of the island itself. While the long-term future of the island remains unclear and the drama for its residents is sure to continue, one thing has been settled. Dorothy Tresness has won the right to be independent, to remain an island unto herself. Through decades of change, and through changes yet to come, she will continue to make her own place on the island and in its history. Her ultimate victory will continue to play out as she returns to the river each summer, to come home to Bob-Lo Island.